T0311652

STATISTICS IN MATLAB®
A PRIMER

Chapman & Hall/CRC
Computer Science and Data Analysis Series

The interface between the computer and statistical sciences is increasing, as each discipline seeks to harness the power and resources of the other. This series aims to foster the integration between the computer sciences and statistical, numerical, and probabilistic methods by publishing a broad range of reference works, textbooks, and handbooks.

SERIES EDITORS
David Blei, Princeton University
David Madigan, Rutgers University
Marina Meila, University of Washington
Fionn Murtagh, Royal Holloway, University of London

Proposals for the series should be sent directly to one of the series editors above, or submitted to:

Chapman & Hall/CRC
Taylor and Francis Group
3 Park Square, Milton Park
Abingdon, OX14 4RN, UK

Published Titles

Semisupervised Learning for Computational Linguistics
Steven Abney

Visualization and Verbalization of Data
Jörg Blasius and Michael Greenacre

Design and Modeling for Computer Experiments
Kai-Tai Fang, Runze Li, and Agus Sudjianto

Microarray Image Analysis: An Algorithmic Approach
Karl Fraser, Zidong Wang, and Xiaohui Liu

R Programming for Bioinformatics
Robert Gentleman

Exploratory Multivariate Analysis by Example Using R
François Husson, Sébastien Lê, and Jérôme Pagès

Bayesian Artificial Intelligence, Second Edition
Kevin B. Korb and Ann E. Nicholson

Published Titles cont.

Computational Statistics Handbook with MATLAB®, Second Edition
Wendy L. Martinez and Angel R. Martinez

Exploratory Data Analysis with MATLAB®, Second Edition
Wendy L. Martinez, Angel R. Martinez, and Jeffrey L. Solka

Statistics in MATLAB®: A Primer
Wendy L. Martinez and MoonJung Cho

Clustering for Data Mining: A Data Recovery Approach, Second Edition
Boris Mirkin

Introduction to Machine Learning and Bioinformatics
Sushmita Mitra, Sujay Datta, Theodore Perkins, and George Michailidis

Introduction to Data Technologies
Paul Murrell

R Graphics
Paul Murrell

Correspondence Analysis and Data Coding with Java and R
Fionn Murtagh

Pattern Recognition Algorithms for Data Mining
Sankar K. Pal and Pabitra Mitra

Statistical Computing with R
Maria L. Rizzo

Statistical Learning and Data Science
Mireille Gettler Summa, Léon Bottou, Bernard Goldfarb, Fionn Murtagh, Catherine Pardoux, and Myriam Touati

Foundations of Statistical Algorithms: With References to R Packages
Claus Weihs, Olaf Mersmann, and Uwe Ligges

Computer Science and Data Analysis Series

STATISTICS IN MATLAB®
A PRIMER

WENDY L. MARTINEZ

BUREAU OF LABOR STATISTICS
WASHINGTON, D.C., USA

MOONJUNG CHO

BUREAU OF LABOR STATISTICS
WASHINGTON, D.C., USA

CRC Press
Taylor & Francis Group
Boca Raton London New York

CRC Press is an imprint of the
Taylor & Francis Group, an **informa** business

A CHAPMAN & HALL BOOK

CRC Press
Taylor & Francis Group
6000 Broken Sound Parkway NW, Suite 300
Boca Raton, FL 33487-2742

© 2015 by Taylor & Francis Group, LLC
CRC Press is an imprint of Taylor & Francis Group, an Informa business

No claim to original U.S. Government works

Printed on acid-free paper
Version Date: 20141113

International Standard Book Number-13: 978-1-4665-9656-6 (Paperback)

Visit the Taylor & Francis Web site at
http://www.taylorandfrancis.com

and the CRC Press Web site at
http://www.crcpress.com

*Wendy dedicates this book to her parents
who started her on this path:*

Shirley and Glenn Cukr

MoonJung dedicates this book to her children:

Catherine and Ted

Table of Contents

Chapter 3
Descriptive Statistics

Chapter 4
Probability Distributions

List of Figures

Chapter 1
MATLAB® Basics

Chapter 2
Visualizing Data

Chapter 3
Descriptive Statistics

Chapter 4
Probability Distributions

Chapter 5
Hypothesis Testing

Chapter 6
Model-Building with Regression Analysis

Chapter 7
Multivariate Analysis

Chapter 8
Classification and Clustering

List of Tables

Preface

The functionality in MATLAB® for statistical data analysis has improved and expanded in the past several years, with major changes made in 2012 (version 8). Additionally, MATLAB is frequently used in academia to teach statistics, engineering, mathematics, and data analysis courses. Thus, we felt that a book that provides an overview or introduction to the extensive functionality available in MATLAB would be useful to a wide audience.

The main MATLAB software includes many basic functions for statistical visualization and data analysis. The MathWorks, Inc. Statistics Toolbox extends these basic capabilities by including additional specialized functions. The Statistics Toolbox can be purchased separately. MathWorks also has a Student Version of MATLAB that includes the Statistics Toolbox, as well as many other toolboxes.

One should have the Statistics Toolbox to get the most from this book. However, we include enough content to help those who do not have the extra MATLAB functionality. We have been careful to note where to find the functions—in the base MATLAB or the Statistics Toolbox. If the reader is ever confused about where the function comes from, type

which *functionname*

at the command line, and the location of the function file will be displayed.

For example, **normpdf** is a function in the Statistics Toolbox. We get the following, when we type **which normpdf**:

C:\MATLAB2013a\toolbox\stats\stats\normpdf.m

This is under the directory **~\toolbox\stats**, indicating the function is part of the Statistics Toolbox. A function that is in base MATLAB will have a directory **~\toolbox\matlab**.

It took over a year to write this book and several versions of MATLAB. We started with MATLAB R2013a and finished with MATLAB 2014a. All functions should work with R2013, with a few exceptions. These are found mostly in Chapter 8. We recommend that readers investigate the changes in MATLAB by looking at the release notes that are available in the MATLAB documentation or try to have the latest version installed.

There is a companion website, where the reader can find the data sets, M-files with the code from the book, and additional examples. This website is

www.pi-sigma.info

You can also download the files from the book website at CRC Press. For the most part, we assume the reader has a basic knowledge of statistics and probability. The focus of this book is on how to use the statistics capabilities in MATLAB, not on the theory and use of statistics. However, we do include definitions and formulas in certain areas in order to aid the understanding of the MATLAB functions.

The reader should also know some basic concepts of linear algebra. This includes the definitions of vectors, matrices and operations such as adding vectors, multiplying matrices, taking transposes, etc.

Please note that this book is an *introduction*. It is not meant to cover all aspects of statistical analysis nor all of the functions available for statistics in MATLAB. Readers should always refer to the help files and other documentation provided with MATLAB to get the full story. We provide information on how to access the documentation in the first chapter.

We would like to acknowledge the invaluable help of the reviewers: Tom Lane, John Eltinge, Terrance Savitsky, Eungchun Cho, Ted Cho, and Angel Martinez. Their many helpful comments and suggestions resulted in a better book. Any shortcomings are the sole responsibility of the authors. We greatly appreciate the help and patience of those at CRC Press: David Grubbs, Jessica Vakili, Robin Starkes, and Kevin Craig. Finally, we are indebted to Naomi Fernandes, Paul Pilotte, and Tom Lane at The MathWorks, Inc. for their special assistance with MATLAB.

Disclaimers

1. Any MATLAB programs and data sets that are included with the book are provided in good faith. The authors, publishers, or distributors do not guarantee their accuracy and are not responsible for the consequences of their use.

2. MATLAB® and Simulink® are trademarks of the MathWorks, Inc. and are used with permission. The MathWorks does not warrant the accuracy of the text or the exercises in this book. This book's use or discussion of MATLAB® and Simulink® software or related products does not constitute endorsement or sponsorship by the MathWorks of a particular pedagogical approach or particular use of the MATLAB® and Simulink® software.

3. The views expressed in this book are those of the authors and do not necessarily represent the views of the United States Department of Labor or its components.

Wendy Martinez
MoonJung Cho

Chapter 1

MATLAB® Basics

Computers are like Old Testament gods; lots of rules and no mercy.
  Joseph Campbell (1904–1987)

The purpose of this chapter is to provide some introductory information that will help you get started using MATLAB. We will describe:

- The desktop environment
- How to get help from several sources
- Ways to get your data into and out of MATLAB
- The different data types in MATLAB
- How to work with arrays
- Functions and commands

We do not have to concern ourselves with a lot of detailed notation because this book will not contain many equations. However, we do use certain fonts to indicate MATLAB functionality, as outlined here.

- MATLAB code will be indicated by Courier bold font: **matlab**.
- Menu, button, and ribbon options will be in bold, small caps: **MENU**.

1.1 Desktop Environment

This section will provide information on the desktop environment to include the following components:

- Toolstrip ribbon interface
- Desktop layouts
- M-File or Script Editor

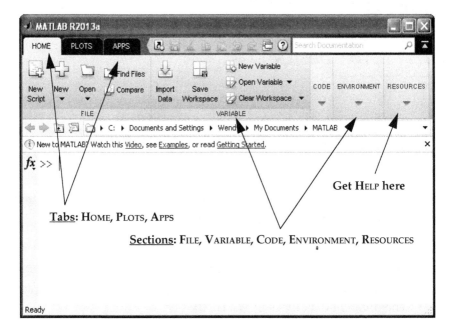

FIGURE 1.1
This is a screenshot of the main MATLAB window or desktop, where we have chosen to view only the Command Window in our layout. The desktop layout can be changed by clicking on the ENVIRONMENT *section and the* LAYOUT *option. MATLAB now uses the familiar ribbon interface common to Microsoft® applications.*

Figure 1.1 shows the desktop layout for MATLAB, where we chose to view just the Command Window. We made the window smaller, which causes some of the sections—CODE, ENVIRONMENT, and RESOURCES, in this case—to be collapsed. Simply click on the arrow buttons for the sections to see what tools are available or make the window bigger by resizing with the mouse.

The default desktop layout includes the following panels or windows:

- Command Window
- Current Folder
- Workspace
- Command History

The default layout will automatically resize the desktop to show all of the sections in the toolstrip ribbon.

You can personalize your desktop layout by choosing those panels or subwindows that you need for your work, dragging the panels to different places within the desktop, or re-sizing them. We now describe the commonly used components of the desktop environment in more detail.

Command Window

The Command Window is the main interface between you and MATLAB. The window shows the MATLAB prompt, which is shown as a double arrow: >>. This is where you enter commands, functions, and other code. MATLAB will also display information in this spot in response to certain commands.

Toolstrip Ribbon and Menus

MATLAB now uses the typical ribbon interface found in the recent Microsoft Windows® applications. The interface for MATLAB is comprised of tabs and sections, as highlighted in Figure 1.1. The ribbon interface will be different and depends on the selected tab (HOME, PLOTS, or APPS).

We will provide more detail on the Figure window in a later chapter, but we mention it here briefly. Plots are displayed in a separate Figure window with its own user interface, which is comprised of the more familiar menu options and toolbar buttons. The main menu items include FILE, EDIT, TOOLS, HELP, and more.

Workspace

The Workspace sub-window provides a listing of the variables that are in the current workspace, along with information about the type of variable, the size, and summary information. You can double-click on the VARIABLE icon to open it in a spreadsheet-like interface. Using this additional interface, you can change elements, delete and insert rows, select and plot columns, and much more.

Script Editor

We will not discuss programming in this book, but the general user will find the Script Editor to be a useful tool. This can be opened by clicking the NEW SCRIPT button on the ribbon interface. This editor has its own ribbon interface with helpful options for programming in MATLAB. The editor also has some nice features to make it easier to write error-free MATLAB code, such as tracking parentheses, suggesting the use of a semi-colon, and warnings about erroneous expressions or syntax.

TIP

You can separate the panels (or sub-windows) from the desktop. Click on the arrow in the upper right corner of the panel and select the UNDOCK menu item. Select DOCK to put it back onto the desktop.

1.2 Getting Help and Other Documentation

This book is meant to be a brief introduction to statistics and data analysis with MATLAB—just enough to get you started. So, the reader is encouraged to make use of the many sources of assistance that are available via the online documentation, command line tools, help files, and the user community. One obvious way to start looking for these sources of help is to select the HELP button that can be found on the HOME ribbon and the RESOURCES section.

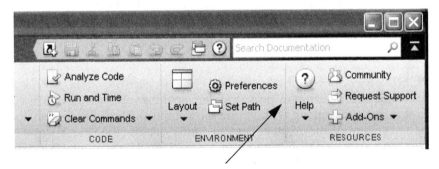

FIGURE 1.2
This shows the options available from the HELP *button on the* RESOURCES *section of the* HOME *tab. Click the question mark to open a documentation window or click the arrow button to access more options.*

Command Line Options

There are several options that you can use to get help on MATLAB functions through the command line interface. They are the easiest and quickest way to get some help, especially once you become familiar with MATLAB.

You can find a shortened version of the documentation for a function by typing

$$\texttt{help } \textit{functionname}$$

at the command line. This will provide information on the syntax for the function, definitions of input/output arguments, and examples. The command

doc *functionname*

will open the documentation for *functionname* in a separate window.

You can get tooltip help that will display the syntax portion of the help file when you start typing the function name and pause after typing the opening parentheses, as shown here

functionname(

Of course, these options assume you know the name of the function. MATLAB provides a function called **lookfor** that can be used in cases where you do not know the name, but you have some idea of what the function does or some keyword associated with it. For example, we know there is a function in the Statistics Toolbox that will construct parallel coordinate plots, but we cannot remember the exact function name to get help. We can conduct a search by using

lookfor parallel,

which searches the first line of the help text in MATLAB files. We get the following results from our search:

```
>> lookfor parallel
...
parallelcoords - Parallel coordinates plot
...
```

Now, we could type **help parallelcoords** to get the syntax and other information.

TIP

Using **help** *functionname* will return links to the documentation page and related functions. Using **help** alone returns links to help topics.

Documentation

You can access the documentation in several ways. One is via the HELP button (see Figure 1.2) on the HOME ribbon. This opens a window that has links to the documentation for your installed toolboxes. There is also a link to .pdf documentation on all toolboxes at the bottom of this list. Clicking on that link will take you to the Documentation Center at the MathWorks, Inc. website.

User Community

There is a vast user community for MATLAB, and this is a great resource for finding answers, obtaining user-written code, and following news groups.

The main portal for this community is MATLAB Central, and it can be found at this link:

http://www.mathworks.com/matlabcentral/

1.3 Data Import and Export

Getting data into MATLAB is the first step in any analysis. Similarly, we also need to have the ability to export data for use in other software (such as Microsoft® Excel®, SPSS, R, etc.) or to save the objects we created for future analysis. MATLAB has many ways to import and export data (***data I/O***) that are in various file formats, and they fall into two approaches: using command line functions or the interactive Import Wizard.

In this section, we show you how to import and export data using these two main approaches that are available with the base MATLAB software. We also discuss the different types of data files that can be easily imported into MATLAB for your analysis. These include ASCII text files, character-delimited formats, Microsoft Excel spreadsheets, tabular data and the MATLAB specific format with the **.mat** extension. Next, we discuss how you can export data from MATLAB and save it in similar file formats. The section concludes with some examples using the functionality we describe and a list of additional data I/O options for specialized types of data, such as images, audio, and more.

1.3.1 Data I/O via the Command Line

.mat *Files*

The main functions that can be used for importing and exporting MATLAB specific data files via the command line are the **load** function for importing and the **save** function used for exporting. These functions can also be used with an ASCII text file.

Text Files

MATLAB has several options for reading in ASCII text files. If the file has numerical entries separated by white space, then the **load** (for exporting use **save -ascii**) command can be employed. If the values are separated by a different character, such as commas or tabs, then the **dlmread** is a good option for loading the data into MATLAB. The character used to separate the values is inferred from the file or it can be specified as an argument. A related function called **dlmwrite** is used for exporting data in an ASCII file with different delimiters.

Comma-Separated Value (**.csv**) Files

Most software packages for data analysis provide an option to import and export files where the data entries are separated by commas. MATLAB provides the functions **csvread** and **csvwrite** for this common type of file. Note that the file must contain only numeric data to use these functions.

Spreadsheets

You can import and export spreadsheet files with the **.xls** or **.xlsx** extensions using **xlsread** and **xlswrite**. The **xlsread** function will also read in OpenDocument™ spreadsheets that have the **.ods** file extension. For more information on OASIS OpenDocument formats, see

http://www.opendocumentformat.org/.

Files with Characters and Numbers

Sometimes a file can have numerical data, as well as some text. For example, the file might have non-numeric data like formatted dates, along with some numeric variables. There are two command-line options for reading in these types of files. One is the **importdata** function. This function can handle headers for ASCII files and spreadsheets, but the rest of the data should be in tabular form, and it has to be numeric. If you have an ASCII file with columns of non-numeric data (characters or formatted dates/times), then you can use the **textscan** function or the Import Wizard described next.

TABLE 1.1

Common Data I/O Functions in Base MATLAB®

load, save	Read and write **.mat** files Read and write text files using the **-ascii** flag
dlmread, dlmwrite	Handles text files with specified delimiter
csvread, csvwrite	Use for comma value or **.csv** files
xlsread, xlswrite	Read and write spreadsheet files
importdata, textscan	Use for files that have a mixture of text and numbers

1.3.2 The Import Wizard

You can invoke the Import Wizard by clicking on the IMPORT DATA button that is located on the MATLAB desktop HOME ribbon's VARIABLE section. You can also start it by typing in **uiimport** at the command line. The wizard guides you through the process of importing data from many different recognized file types, such as the ones we have already discussed. There are also options for reading images, audio, and video data into MATLAB. It allows you to view the contents of a file, to select the variables and the observations to import, to specify a value for un-importable cells, and more.

FIGURE 1.3
Select the IMPORT DATA *button to start the Import Wizard.*

1.3.3 Examples of Data I/O in MATLAB®

We now provide some examples of using the **load** and **save** commands and the Import Wizard. To help us get started, we can view what is in our current directory by typing **dir** at the command line:

```
>> dir
.                 DensityEarth.txt iris.mat
..                UStemps.txt
```

These are the files we will be working with in the following examples. You can download all of the data sets from the book website

Density of the Earth

The **DensityEarth.txt** file contains a data set comprised of 29 measurements of the density of the earth, given as a multiple of the density of water [Stigler, 1977]. These data were downloaded from the *Data and Story Library (DASL)*, which is a repository of data sets and stories that help explain basic methods in statistics. A link to the site is given here

http://lib.stat.cmu.edu/DASL/.

Iris Data

The data in the **iris.mat** file are commonly known as Fisher's iris data. They were originally collected by Anderson in 1935 [Anderson, 1935] and were subsequently analyzed by Fisher in 1936 [Fisher, 1936]. The data set consists of 150 observations, each containing four measurements—sepal length and width, petal length and width—in centimeters. These were collected for three species of iris: *Iris setosa*, *Iris virginica*, and *Iris versicolor*. The goal of the original study was to develop a methodology that could be used to classify an iris as belonging to one of the three species based on these four variables.

US Temperatures

These data were downloaded from the DASL website. The data set contains 56 observations and the following variables:

- **City**: City and state abbreviation
- **JanTemp**: Average temperature (degrees Fahrenheit) for the month of January during the years 1931 to 1960
- **Lat**: Latitude in degrees north of the equator
- **Long**: Longitude in degrees west of the prime meridian

The first variable (**City**) is text or characters, and the remaining three variables are numeric. The data are in a file called **UStemps.txt**.

TIP

You can put comments in your MATLAB code by using the percent (%) sign. Words after the % sign are ignored.

Loading and Saving **.mat** *Files*

We are first going to load the file called **iris.mat**. The file **iris.mat** can be downloaded from the book website.

```
% First see what is in the workspace. This command
% lists all variables in the workspace.
who

% The workspace is empty and nothing is returned.
% Now load the iris.mat file.
load iris

% What is in the workspace now?
who
```

```
% This is what we see in the command window:
Your variables are:

setosa versicolor virginica

% Now save the variables in another file.
% We will save just the setosa variable object.
% Use save filename varname.
save setosa setosa

% See what files are in the current directory.
dir
% This is what you see in the command window:

.              DensityEarth.txt   iris.mat
..             UStemps.txt        setosa.mat
```

TIP

To export *all* variables to a .mat file use the command **save** *filename*.

Loading and Saving Text Files

In this example, we will show how to load and save text files. A **.mat** file can have many variables (or objects), but a text file can have only one data object. When loaded, the object is saved in the workspace with the same name as the text file.

```
% Remove objects from workspace to clean it up.
clear

% The directory should be empty. Load the earth data.
load DensityEarth.txt -ascii

% See what is in the workspace.
who

% This is what we see in the command window:
Your variables are:

DensityEarth
```

We first used the **clear** command to remove all objects from the workspace, just to clean things up. Note that we have an object called **DensityEarth**, after we loaded the file. To save an object (or variable) in an ascii text file, use a similar syntax: **save** *filename varname* **-ascii**.

Import Wizard

The Import Wizard is an intuitive graphical user interface (GUI) that will guide the user through the import process. It can be used with many file types. We will use it to import the **UStemps.txt** file. Here are the steps.

1. Click the IMPORT DATA button on the VARIABLE section of the HOME ribbon (see Figure 1.3).
2. Select the file you want to open and import.
3. A window will come up that displays the variable names (if given in the file) in the column headers, cells with the values, and other useful information (see Figure 1.4).
4. Make any desired changes or selections and click on the IMPORT SELECTION to save the objects (or variables) to the workspace.

You will get a small pop-up window that tells you what variables were saved to the workspace. We followed this procedure with the **UStemps.txt** file. We can check the workspace to see what objects we have now.

```
% Remove objects for simplicity.
clear

% See what is in the workspace.
who

% This is displayed in the command window.
Your variables are:

City     JanTemp  Lat     Long
```

1.3.4 Data I/O with the Statistics Toolbox

There are some additional command line functions for reading in and exporting data that come with the Statistics Toolbox; see Table 1.2 for a summary of the functions. These are particularly useful for statisticians and data analysts, and they include functions for importing and exporting tabular data, as well as data stored in the SAS XPORT format.

Tabular Data:

The functions **tblread** and **tblwrite** will import and export data that are in a tabular or matrix-like format. The data file must have the variable names on the first row, and the case names (or record identification variable) are in the first column. The data entries would start in the cell corresponding to the second row and second column (position $(2, 2)$ in a matrix).

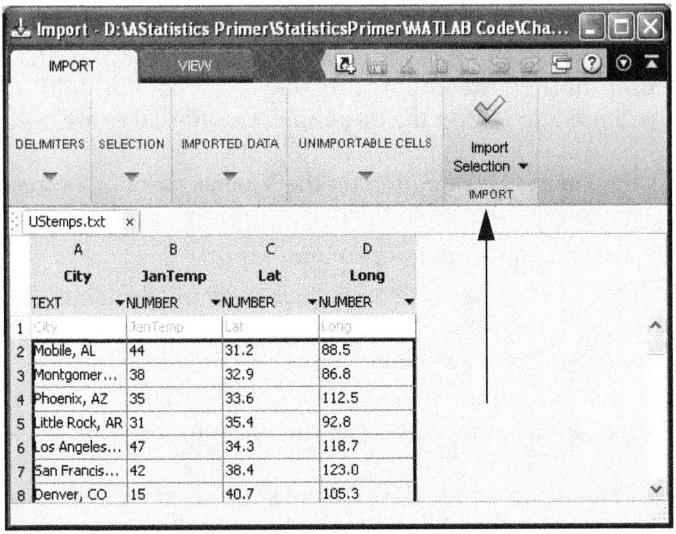

FIGURE 1.4
You should have a similar window when you use the Import Wizard to load the **UStemps.txt** *file. Click on the* IMPORT SELECTION *button once you make your selection and changes. The selected variables will be saved to the workspace using the names designated in each column. You can click on the heading names in this view to edit them.*

The basic syntax to *interactively* select a file to load is:

[data, vnames, cnames] = tblread,

where **vnames** contains the variable names (first row of the file) and **cnames** has the names of the observations (first column of the file). The variable **data** is a numeric matrix, where rows correspond to the observations, and the columns are the variables or characteristics.

Calling **tblread** with no input argument opens a window interface for selecting files. You can also specify a file name by using an input argument, as follows:

[data, vnames, cnames] = tblread(*filename*).

The following function call

tblwrite(data,vnames,cnames,*filename*,delim)

will export the data to a file with the delimiter specified by **delim**. Use **help tblwrite** at the command line for delimiter options.

SAS Files

Statisticians and data analysts often have to read in SAS files. The function **xptread** can be used to import files that are in the SAS XPORT transport format. The file can be selected interactively by leaving out the file name as the input argument or you can use the following syntax to read in a specific file:

$$data\ =\ xptread(\textit{filename}).$$

TABLE 1.2

Data I/O Functions in the Statistics Toolbox

tblread, tblwrite	Data in tabular format with variable names on the first row and case names in the first column
xptread	SAS XPORT (transport) format files
caseread, casewrite	Import and export text files with one case name per line
export	Write a dataset array to a tab-delimited file
tdfread	Import a tab-delimited file that has text and numeric data

1.3.5 More Functions for Data I/O

MATLAB has more functionality for importing and exporting data than what we described above. This includes functions for special files like images, scientific data, Web content, and more. Some of the additional file formats that can be handled with the base MATLAB software are summarized in Table 1.3. See the **help** files on the individual functions and accompanying documentation for more details.

TABLE 1.3

Advanced I/O Functions in MATLAB®

imread, imwrite	Common image formats: BMP, GIF, JPEG, PNG
audioread, audiowrite	Common audio formats: WAV, MP3, FLAC
xmlread, xmlwrite	XML documents
urlread, urlwrite	Contents of Web pages
fscanf, fprintf	Low-level file I/O for formatted data

It is always a good idea to check any toolboxes you might have for data I/O functions that are unique to that area. For example, the documentation for the Statistics Toolbox describes the functions we identified above and some additional ones for reading and exporting case names (`caseread`, `casewrite`). MATLAB documentation and help pages can be accessed from the desktop environment by clicking on the HELP button (on the HOME ribbon), selecting STATISTICS TOOLBOX ... EXPLORATORY DATA ANALYSIS ... DATA IMPORT AND EXPORT.

1.4 Data in MATLAB®

In this section, we describe the basic data types in MATLAB and the Statistics Toolbox. We also discuss how to merge data sets and to create special arrays that might prove useful in data analysis. We conclude the section with a short introduction to object-oriented programming constructs and how they are used in MATLAB.

1.4.1 Data Objects in Base MATLAB®

One can consider two main aspects of a data object in MATLAB—the object class and what it contains. We can think of the object *class* as the type of container that holds the data. The most common ones are arrays, cell arrays, and structures. The *content* of objects in MATLAB can be numeric (e.g., double precision floating point or integers) or characters (e.g., text or strings). We now describe the common types or classes of objects in the base MATLAB software.

Arrays

The word *array* is a general term that encompasses scalars, vectors, matrices, and multi-dimensional arrays. All of these objects have a *dimension* associated with them. You can think of the dimension as representing the number of indexes you need to specify to access elements in the array. We will represent this dimension with the letter k.

- **Scalar**: A scalar is just a single number (or character), and it has dimension $k = 0$. You do not need to specify an address because it is just a single element.

- **Vector**: A vector is usually a column of values, but MATLAB also has row vectors. A vector has dimension $k = 1$, because you have to provide one value to address an element in the column (or row) vector.

- **Matrix**: A matrix is an object that has rows and columns—like a table. To extract or access an element in a matrix, you have to specify what row it is in and also the column. Thus, the dimension of a matrix is $k = 2$.

- **Multi-dimensional array**: A multi-dimensional array has dimension $k > 2$. For example, we could think of a three-dimensional array being organized in pages (the third dimension), where each page contains a matrix. So, to access an element in such an array, we need to provide the row, column, and page number.

Building Numeric Arrays

As a data analyst, you will usually import data using **load** or some other method we described previously. However, you will likely also need to type in or construct arrays for testing code, entering parameters, or getting the arrays into the right form for calling functions. We now cover some of the ways to build small arrays, which can also be used to combine separate arrays, as we illustrate in the examples.

Commas or spaces concatenate elements (or other arrays) as columns. In other words, it puts them together as a row. Thus, the following MATLAB code will produce a row vector

$$x = [1, 4, 5]$$

Or, we can concatenate two column vectors **a** and **b** to create one matrix, as shown here

$$Y = [a\ b]$$

The semi-colon will stack elements as rows. So, we would obtain a column vector from this command:

$$z = [1;\ 4;\ 5]$$

As another example, we could put three row vectors together to get a matrix, as shown below

$$Y = [a;\ b;\ c]$$

It is important to note that the building blocks of your arrays have to be conformal in terms of the number of elements in the sub-arrays when using the comma or semi-colon to merge data sets. Otherwise, you will get an error.

We might need to generate a regular sequence of values when working in MATLAB. We can accomplish this by using the colon. We get a sequence of values from one to ten with the following syntax

$$x = 1:10$$

Other step sizes can be used, too. For instance, this will give us a sequence from one to ten in steps of 0.5:

$$x = 1:0.5:10$$

and this will yield a decreasing sequence:

$$x = 10:-1:1$$

We can create an array of all zeros or all ones. Arrays of this type are often used in data analysis. Here is how we can create a 3×3 matrix of zeros:

$$Z = \text{zeros}(3,3)$$

This next function call will produce a multidimensional array with $k = 3$:

$$O = \text{ones}(2,4,3)$$

The array **O** has three pages (third dimension), and each page has a matrix with two rows and four columns.

> _TIP_
>
> Create a string array using quotes: `strg = 'This is a string'`

Empty Arrays

There is a special type of array in base MATLAB that data analysts will find useful. This is called the *empty array*. An empty array is one that contains no elements. Thus, it does not have any dimensions associated with it, as we mentioned above with other arrays. The empty array is designated by closed square brackets, as shown here: `[]`. It can be used to delete elements of an array at the command line, and it is sometimes returned in response to function calls and logical expressions.

Here is an example of the first case, where we show how to delete an element from a vector **x**.

```
% Create a vector x.
x = [2, 4, 6];

% Delete the second element.
x(2) = [];

% Display the vector x.
disp(x)
       2       6
```

Next, we provide an example where we get an empty array in response to a function call and a logical expression.

```
% Find the elements that are negative.
ind = find(x < 0);
```

```
% Print the vector ind to the command window.
ind

% This is displayed:
ind =

    Empty matrix: 1-by-0
```

Cell Arrays

A *cell array* is an object type in the base MATLAB software, and later on, we will see that these are particularly useful when working with strings. The elements of a cell array are called *cells*. Each cell provides a flexible container for our data, because it can hold data of any type—even other cell arrays. Furthermore, each element of the cell array can have a different size.

The cell array has an overall structure that is similar to basic numeric or character data arrays covered previously, and as such, they have to be conformal in their overall arrangement. For example, the cells are arranged in rows, columns, pages, etc., as with arrays. If we have a cell array with two rows, then each of its rows has to have the same number of cells. However, the content of the cells can be different in terms of data type and size. One cell might contain a vector of ten character elements, another might have twenty numeric values arranged in a matrix, and some could be empty.

We can create an empty cell array using the function **cell**, as shown here, where we set up a $2 \times 4 \times 3$ array of cells. Each of the cells in the **cell_arry** object is empty.

```
cell_arry = cell(2,4,3)
```

You can also construct a cell array and fill it with data, which is shown in the code below.

```
% Create a cell array, where one cell contains
% numbers and another cell element is a string.
cell_arry2 = {[1,2], 'This is a string'};

% Let's check the size of the cell array
size(cell_arry2)

ans =

        1       2
```

Note that curly braces are used to denote a cell array.

Structures

Like cell arrays, *structures* allow one to combine dissimilar data into a single variable. You can think of a structure as being somewhat like a table in a relational database, in that structures have *records* (or observations) and *fields* (or characteristics).

The basic syntax to create a structure is

```
S = struct('field1',data1,'field2',data2,...).
```

Note that the structure can have one or more fields, along with the associated data values. Let's use this to create a small structure.

```
% Create a structure called employee with three fields.
employee = struct(...
    'name',{{'Wendy','MoonJung'}},...
    'area',{{'Visualization','Inference'}},...
    'deg',{{'PhD','PhD'}},...
    'score',[90 100])

% There is no semi-colon at the end.
% So, the following is displayed.

employee =

    name: {'Wendy'    'MoonJung'}
    area: {'Visualization'    'Inference'}
     deg: {'PhD'    'PhD'}
   score: [90 100]
```

A dot notation is used to extract or access a field. Suppose we want to get all of the names in our **employee** structure, then we can use

```
all_names = employee.name
```

to get all of the entries in the **name** field. We will discuss how to access individual data elements and records in the next section.

Tables

A *table* is a type of data object in the base MATLAB software. The table object is like cell arrays and structures, because we can combine data of different data types in one object. However, it has a table-like format that is familiar to statisticians and data analysts. The rows of a table object would contain the observations or cases, and the columns correspond to the characteristics or features.

We can create a table in several ways, and we illustrate one of them below. This uses the **table** function to create a table object employing variables that are in the workspace.

```
load UStemps
% Create a table using all four variables.
UTs_tab = table(City,JanTemp,Lat,Long)

% This appears in the workspace.
UTs_tab = table(City,JanTemp,Lat,Long)

UTs_tab =
```

City	JanTemp	Lat	Long
'Mobile, AL'	44	31.2	88.5
'Montgomery, AL'	38	32.9	86.8
'Phoenix, AZ'	35	33.6	112.5
'Little Rock, AR'	31	35.4	92.8
'Los Angeles, CA'	47	34.3	118.7
'San Francisco, CA'	42	38.4	123

...

View what is in the workspace by typing **whos** at the prompt, and you will get a list of variables as shown here. Note that there is a column that indicates the class of the variable. This shows us that the **City** variable is a cell object, and **JanTemp**, **Lat**, and **Long** are numeric arrays. We were able to put these variables with different data types into one container called **UTs_tab**, and we see that it has a class of table.

```
% This is displayed in the window.
```

Name	Size	Bytes	Class
City	56x1	4790	cell
JanTemp	56x1	448	double
Lat	56x1	448	double
Long	56x1	448	double
UTs_tab	56x4	7492	table

You can import a file as a table object using the **readtable** function. This function works with delimited text files (**.txt**, **.dat**, or **.csv**). It will also read in an Excel spreadsheet file with **.xls** or **.xlsx** extensions.

TIP

Another useful data object in base MATLAB is the **categorical** object. This can be used to hold values from a finite set of discrete classes or categories. The classes can be denoted by numbers or strings.

1.4.2 Accessing Data Elements

In this section, we demonstrate how you can identify elements of arrays, cell objects, and structures. This is useful in data analysis, because we often need to analyze subsets of our data or to create new data sets by combining others.

Array Elements

We mentioned previously that it is possible to access elements of your arrays using the Workspace window when you want to view and/or edit elements of your arrays. However, you will likely need to accomplish this in functions or commands that you enter on the command line. Table 1.4 provides some examples of how to access elements of arrays—both numeric and cell arrays.

The examples in Table 1.4 are for elements of numeric (or string) arrays or cell arrays. This notation is used to access the cell elements, but not the *contents* of the cells. Curly braces **{ }** are used to get to the data that are inside the cells. For example, **A{1,1}** would give us the contents of the cell, which would have a numeric or character type. Whereas, **A(1,1)** is the cell, and it has a type (or class) of cell.

These two notations can be combined to access part of the contents of a cell. To get the first two elements of the vector contents of cell **A(1,1)**, we can use

A{1,1}(1:2)

The curly braces in **A{1,1}** tells MATLAB to go inside the cell in position **(1,1)**, and the **(1:2)** points to elements 1 and 2 inside the cell.

Accessing Structure Elements

Recall that we can access entire fields in a structure using the dot notation, as shown here using our **employee** structure created previously.

```
e_area = employee.area;

% Display the contents in the window.
e_area
```

```
e_area =

    'Visualization'      'Inference'
```

We can extract partial content from the fields by using the techniques we described for numeric and cell arrays, as illustrated below.

```
% Display Wendy's score.
employee.score(1)
```

```
ans = 90
```

TABLE 1.4

Examples of Accessing Elements of Arrays

Notation	Usage
`a(i)`	Access the *i*th element (cell) of a row or column vector array (cell array)
`a(3:5)`	Access elements 3 through 5 of a vector or cell array
`A(:,i)`	Access the *i*th column of a matrix or cell array. In this case, the colon in the row dimension tells MATLAB to access all rows.
`A(i,:)`	Access the *i*th row of a matrix or cell array. The colon tells MATLAB to gather all of the columns.
`A(2:4,1:2)`	Access the elements in the second, third, and fourth rows and the first two columns
`A(1,3,4)`	Access the element in the first row, third column on the fourth entry of dimension 3 (sometimes called the page).

The **area** field is a cell array of strings. We use the curly braces to get the contents of the cell, as shown here.

```
% Get MoonJung's area.
employee.area{2}
```

```
ans = Inference
```

Note that we have to employ the dot notation to first access the field and then specify the elements we want using the usual notation for arrays.

Accessing Table Elements

The techniques for manipulating subsets of data in table objects are similar to structures and arrays, but they have some additional options because of the column or variable names.

We illustrate some of the approaches for accessing portions of tables using the table object we created using the US temperature data. This shows how to create a sub-table with the first three records and all variables.

```
% Get a partial table by extracting the first
% three rows.
U1 = UTs_tab(1:3,:)
```

```
% This should be displayed.
U1 =
```

	City	JanTemp	Lat	Long
	'Mobile, AL'	44	31.2	88.5
	'Montgomery, AL'	38	32.9	86.8
	'Phoenix, AZ'	35	33.6	112.5

We are able to extract a column of the table using the dot notation that we had with structures.

```
% Get the JanTemp variable.
jt = UTs_tab.JanTemp;
```

Let's see if it matches the contents of the **JanTemp** vector, which should be in the workspace.

```
% See if it is equal to the JanTemp variable.
isequal(jt,JanTemp)

ans =     1

% We get an answer of 1, indicating they are the same.
```

We can also get a subset of the variables for a given set of records, as shown here.

```
% We can extract the Lat and Long data for the first
% three cities using the variable names.
U2 = UTs_tab{1:3,{'Lat','Long'}}

% This is displayed in the window.
U2 =

    31.2000    88.5000
    32.9000    86.8000
    33.6000   112.5000
```

1.4.3 Examples of Joining Data Sets

In our experience, one of the first tasks in data analysis is to get the data ready for the study. This often requires one to join or combine data in various ways. We now illustrate how you can combine data sets—either horizontally or vertically—using the data that we loaded earlier in the chapter.

Stacking Arrays

Recall that the **iris** data set had three variables: **setosa, virginica**, and **versicolor**. We need to combine these three objects into one data matrix in order to use some functions in MATLAB. For example, we might want to partition the data using the function **kmeans** that will group the data into k clusters using the k-means approach, and this requires a single data matrix. See Chapter 8 for details.

So, using the **iris** data we can stack the variables as follows:

```
% Assumes the iris.mat file is already loaded into
% the workspace. Now, we use the semicolon to stack
% setosa, versicolor, and virginica data objects.
irisAll = [setosa; versicolor; virginica];

% Now look at the workspace to see what is there now.
who

% Something like this should appear in the window.

Your variables are:

irisAll      setosa      versicolor  virginica
```

We can check on the dimensions of our new variable **irisAll** using the **size** function.

```
% Check on the size of irisAll.
size(irisAll)

ans =

    150     4
```

Recalling that each species of iris has 50 observations, we see that the species of iris have been correctly combined into the **irisAll** array. It has 150 observations ($3 \cdot 50 = 150$ rows) and 4 columns for the characteristics of sepal length/width and petal length/width.

Joining Along a Row

The next example shows how to glue data together horizontally or along a row. We will use the **UStemps** data for this illustration. These data had four vectors, each one saved in our workspace with a different variable name, as we saw in a previous section. The following MATLAB code shows how to concatenate the numeric variables into one matrix.

```
% Load the data if not in the workspace.
load UStemps

% Use commas to concatenate as a row.
UStemps = [JanTemp, Lat, Long];

% Check the workspace.
who

% These variables should be displayed.
Your variables are:

City      JanTemp  Lat        Long       UStemps

% Verify the size of UStemps:
size(UStemps)

ans =

    56      3
```

We see that **UStemps** now has the variables in one array.

By using the comma and semi-colon properly, you can combine both of these steps in one command, should you need to do so. Here is an example of combining arrays **A**, **B**, **C**, and **D**—first gluing them side by side with the comma and then stacking them as a column using the semi-colon.

$$X = [[A, B] ; [C, D]];$$

TIP

The semi-colon after a statement stops MATLAB from printing the results of the expression to the command line.

1.4.4 Data Types in the Statistics Toolbox

We now give a brief introduction to an additional data object that is defined in the Statistics Toolbox. Like the table object, this can be used to create one object with heterogeneous data types. This is called a dataset array.

Dataset Array

A *dataset array* is a special data object that is included in the Statistics Toolbox, and it can be used to store variables of different data types. As an example, you can combine numeric, logical, character, and categorical data in one array. Each row of the dataset array corresponds to an observation, and each column corresponds to a variable. Therefore, each column has to have elements that are of the same data type. However, the individual columns can be different. For instance, one column can be numeric, and another can be text.

A dataset array can be created using variables that exist in the workspace or as a result of importing data. The function **dataset** is used in either case, as shown here. The basic syntax for creating a dataset array from various file types is shown below.

```
% Create from a tab-delimited text file.
ds = dataset('File','filename.txt')

% Create from a .csv file.
ds = dataset('File','filename.csv','Delimiter',',')

% Create from an Excel file.
ds = dataset('XLSFile','filename.xlsx')
```

There are several options for creating a dataset array from variables in the workspace. They are listed here.

```
% Create by combining three different variables.
ds = dataset(var1, var2, var3);

% Create by converting a numeric matrix called data.
ds = mat2dataset(data);
```

A dataset array has its own set of defined operations, and you cannot operate on this type of array in the same manner as a numeric array. We will discuss this idea in more detail shortly, when we cover object-oriented programming. We will also see examples of dataset arrays in Chapter 6, where we describe some approaches for regression analysis in the Statistics Toolbox.

The dataset array might be removed in future versions of MATLAB, but it is still available in version 2014a. Because of this, it is recommended that you use the **table** object in base MATLAB instead of a dataset array.

1.4.5 Object-Oriented Programming

Certain aspects of MATLAB are *object-oriented*, which is a programming approach based on three main ideas:

1. **Classes and objects**: A *class* is a description or definition of a programming construct. An *object* is a specific instance of a class.
2. **Properties**: These are aspects of the object that can be manipulated or extracted.

3. **Methods**: These are behaviors, operations, or functions that are defined for the class.

The benefit of object-oriented programming is that the computer code for the class and the methods are defined once, and the same method can be applied to different instances of the class without worrying about the details.

Actually, every data object has a class associated with it, and we can view the class of a variable in our workspace in several ways. First, we can use the command **whos** to list all objects in the workspace, along with some other useful information, such as the size in bytes, dimension, and class.

There is a function called **class** that will return the class of an object. This can be very helpful when trying to understand how to access elements and to perform other data analytic tasks. Let's explore these ideas using the table object we created from the US temperature data.

```
% What class is UTs_tab?
class(UTs_tab)

% This is displayed in the window.
ans = table
```

It is not surprising that **UTs_tab** is an instance of the table class.

What about the object **U1**, where we extracted several rows?

```
% What class is U1?
class(U1)

ans = table
```

This also belongs to the table class, because we extracted several observations or rows (or a sub-table).

This was not the case for the object **U2** that contained the **Lat** and **Long** for the first three observations. So, we check the class of that object next.

```
% What class is U2?
class(U2)

ans = double
```

The object is a numeric array. The same is true for **UTs_tab.Lat** or any of the other numeric columns of our table, while the column **UTs_tab.City** is of class cell because it contains text.

```
class(UTs_tab.City)

ans = cell
```

We will encounter some special object classes throughout this book. There are several instances of unique classes that are defined in the Statistics

Toolbox. Some examples of these include probability distributions (Chapter 4), models (Chapter 6), and trees (Chapter 8).

1.5 Miscellaneous Topics

In this section, we cover some additional topics you might find helpful when using MATLAB. These include command line functions for managing your workspace and files, punctuation, arithmetic operators, and functions. We have already used these ideas in our example code, so the purpose of this section is to provide a little more information about the concepts.

1.5.1 File and Workspace Management

You can enter MATLAB expressions interactively at the command line or save them in an M-file. This special MATLAB file is used for saving scripts or writing functions. We described the Script Editor in an earlier section, which is a very handy tool for writing and saving MATLAB code.

As stated previously, we will not be discussing how to write your own programs or functions, but you might find it helpful to write *script* M-files. These script files are just text files with the **.m** extension, and they contain any expressions or commands you want to execute. Thus, it is important to know some commands for file management. There are lots of options for directory and file management on the desktop; we briefly mentioned some interactive tools for interfacing with MATLAB in a previous section. Table 1.5 provides some commands to list, view, and delete files.

Variables created in a session (and not deleted) live in the MATLAB *workspace*. You can recall the variable at any time by typing in the variable name with no punctuation at the end. Note that variable names in MATLAB are case sensitive, so **Temp**, **temp**, and **TEMP** are different variables.

As with file management, there are several tools on the desktop to help you manage your workspace. For example, there is a **Variable** section on the desktop ribbon interface that allows you to create a variable, open current variables in a spreadsheet-like interface, save the workspace, and clear it. Some commands to use for workspace management are given in Table 1.6.

1.5.2 Punctuation in MATLAB®

Punctuation is important in any programming language. If you get this wrong, then you will either get errors or your results will not be what you expect. Some of the common punctuation characters used in MATLAB are described in Table 1.7.

TABLE 1.5

File Management Commands

Command	Usage
dir, ls	Shows the files in the present directory
delete *filename*	Deletes *filename*
pwd	Shows the present directory
cd *dir*	Changes the directory. There is also a pop-up menu and button on the desktop that allows the user to change directory, as we show below.
edit *filename*	Brings up *filename* in the editor
type *filename*	Displays the contents of the file in the command window
which *filename*	Displays the path to *filename*. This can help determine whether a file is part of base MATLAB.
what	Lists the **.m** files and **.mat** files that are in the current directory

TABLE 1.6

Commands for Workspace Management

Command	Usage
who	Lists all variables in the workspace.
whos	Lists all variables in the workspace along with the size in bytes, array dimensions, and object type.
clear	Removes all variables from the workspace.
clear x y	Removes variables **x** and **y** from the workspace.

> **TIP**
>
> **MATLAB saves the commands that you enter. There is a Command History window that you can open using the** Layout **menu. Looking at Figure 1.1, you would click on the** Environment **section in the** Home **ribbon/tab and then the** Layout **button. To re-execute commands, highlight the commands (hold the** ctrl **key to select several), right click on the highlighted area, and click on** Evaluate Selection**. You can also use the arrow keys at the command prompt to recall and move through commands.**

TABLE 1.7

List of MATLAB® Punctuation

Punctuation	Usage
%	A percent sign denotes a comment line. Information after the % is ignored.
,	When used to separate commands on a single line, a comma tells MATLAB to display the results of the preceding command. When used to combine elements or arrays, a comma or a blank space groups elements along a row. A comma also has other uses, including separating function arguments and array subscripts.
;	When used after a line of input or between commands on a single line, a semicolon tells MATLAB not to display the results of the preceding command. When used to combine elements or arrays, a semicolon stacks them in a column.
...	Three periods denote the continuation of a statement onto the next line.
:	The colon specifies a range of numbers. For example, 1:10 means the numbers 1 through 10. A colon in an array dimension accesses all elements in that dimension.

1.5.3 Arithmetic Operators

MATLAB has the usual arithmetic operators for addition, subtraction, multiplication, division, and exponentiation. These are designated by +, −, *, /, ^ , respectively. It is important to remember that MATLAB interprets these operators in a linear algebra sense and uses the corresponding operation definition for arrays.

For example, if we multiply two matrices **A** and **B**, then they must be dimensionally correct. In other words, the number of columns of **A** must equal the number of rows of **B**. Similarly, adding and subtracting arrays requires the same number of elements and configuration for the arrays. Thus, you can add or subtract *row* vectors with the same number of elements, but you will get an error of you try to add or subtract a row and column vector, even if they have the same number of elements. See any linear algebra book for more information on these concepts and how the operations are defined with arrays [Strang, 1993].

The division operator is called a *right matrix divide*. It is considered to be approximately equivalent to solving a system of equations, as shown here:

$$\mathbf{A}/\mathbf{B} \approx \mathbf{A} \times \mathtt{inv}(\mathbf{B})$$

These concepts will be covered in Chapter 6.

Addition and subtraction operations are defined element-wise on arrays, as we have in linear algebra. In some cases, we might find it useful to perform other element-by-element operations on arrays. For instance, we might want to square each element of an array or multiply two arrays element-wise.

To do this, we change the notation for the multiplication, division, and exponentiation operators by adding a period before the operator. As an example, we could square each element of **A**, as follows:

$$\mathbf{A} \mathtt{.\char`\^ 2}$$

A summary of these element-by-element operators are given in Table 1.8.

MATLAB follows the usual order of operations that we are familiar with. The precedence can be changed by using parentheses, which follows the same concepts in mathematics and the syntax of most programming languages.

TABLE 1.8

List of Element-by-Element Operators

Operator	Usage
.*	Multiply element by element
./	Divide element by element
.^	Raise each element to a power

1.5.4 Functions in MATLAB®

MATLAB is a powerful computing environment, but one can also view it as a programming language. Most computer programming languages have mini-programs; these are called *functions* in MATLAB. Functions are similar to macros, for those who are familiar with Microsoft Office products.

In most cases, there are two different ways to call or invoke functions in MATLAB: *function syntax* or *command syntax*, as we describe below. What type of syntax to use is up to the user and depends on what you want to accomplish with the code. For example, you would typically use the function syntax option when the output from the function is needed for other tasks.

Function syntax

The function syntax approach works with input arguments and/or output variables. The basic syntax includes the function name and is followed by arguments enclosed in parentheses. Here is an illustration of the syntax:

```
functionname(arg1, ..., argk)
```

The statement above does not return any output from the function to the current workspace. The output of the function can be assigned to one or more output variables that are enclosed in square brackets:

```
[out1,...,outm] = functionname(arg1,...,argk)
```

You do not need the brackets, if you have only one output variable. The number of inputs and outputs you use depends on the definition of the function and what you want to accomplish. Always look at the **help** pages for a function to get information on the definitions of the arguments and the possible outputs. There are many more options for calling functions than what we describe in this book.

Command Syntax

The main difference between command and function syntax is how you designate the input arguments. With command syntax, you specify the function name followed by arguments separated by spaces. There are no parentheses with command syntax. The basic form of a command syntax is shown here:

```
functionname arg1 ... arg2
```

The other main difference with command syntax pertains to the outputs from the function. You cannot obtain any output values with commands; you must use function syntax for that purpose.

Examples of Syntax

The following MATLAB code shows how we can use function or command syntax to save our data to a **.mat** file. This uses the **save** function we learned about in a previous section.

```
% Save the data for the UStemps to a .mat file.
save UStemps City JanTemp Lat Long


% That used the command syntax to call a function.
% Now use the function syntax.
save('USt.mat','City','JanTemp','Lat','Long')

% See what is in our directory now.
dir

.  DensityEarth.txt  UStemps.mat        iris.mat
.. USt.mat           UStemps.txt        setosa.mat
```

Note that the name of the file and the variable names had to be enclosed in quotes with the function syntax, which was unnecessary with the command syntax. Search the documentation for *syntax* for more information.

1.6 Summary and Further Reading

We tried to provide a brief introduction to some MATLAB fundamentals in this chapter. There are many more things to learn so you can take advantage of this powerful software environment. Therefore, we provide a few pointers to alternative sources of information about MATLAB.

The best source is the online documentation for the base MATLAB and the toolboxes. This can be found at

http://www.mathworks.com/help/index.html

Clicking on the MATLAB link brings up a page with several categories, and we list some of the main ones that are relevant to this book.

- *Getting Started with MATLAB*. Tutorials, desktop, working with arrays and functions
- *Language Fundamentals*: Syntax, data types, array indexing
- *Mathematics*: Linear algebra, basic statistics
- *Graphics*: Two- and three-dimensional plotting
- *Data and File Management*: Importing and exporting data, work-space concepts, directories and folders
- *Advanced Software Development*: Object-oriented programming
- *Desktop Environment*: Setting preferences

You can get to the same webpage by clicking the HELP button in the RESOURCES section of the HOME ribbon and selecting the MATLAB link. The first section—Getting Started with MATLAB—has lots of information to help the new user to become familiar with the software.

At the bottom of this Web page is a button that says **PDF DOCUMENTATION**. This takes you to a page with links to several documents with similar topics we listed above. There is even a *MATLAB Primer* that readers of this book might find useful.

For a comprehensive overview of MATLAB, we recommend *Mastering MATLAB* by Hanselman and Littlefield [2011]. There is a *MATLAB Primer* [Davis, 2010] that should also prove useful to the beginner. Although, our hope is that this chapter will provide enough to get the reader started using MATLAB for statistical data analysis.

Chapter 2

Visualizing Data

Numerical quantities focus on expected values, graphical summaries on unexpected values.
  John Tukey (1915–2000)

All of the functions that we describe in this chapter are contained in the base MATLAB® software. Examples of other specialized visualization capabilities from the Statistics Toolbox will be discussed in the relevant chapters. Here, we will show how to use the basic plotting functions for two-dimensional (2-D) and three-dimensional (3-D) data. We then describe scatter plots and conclude with a presentation of the plotting capabilities available in MATLAB graphical user interfaces (GUIs).

2.1 Basic Plot Functions

In this section, we discuss the main functions for plotting in two and three dimensions. We also describe some useful auxiliary functions to add content to the graph. Type **help graph2d** or **help graph3d** for a list of plotting functions in the base MATLAB software.

2.1.1 Plotting 2-D Data

The main function for creating a 2-D plot is called **plot**. When the function **plot** is called, it opens a new Figure window. It scales the axes to fit the limits of the data, and it plots the points, as specified by the arguments to the function. The default is to plot the points and connect them with straight lines. If a Figure window is already available, then it produces the plot in the current Figure window, replacing what is there.
  The main syntax for **plot** is

```
plot(x,y,'color_linestyle_marker')
```

where **x** and **y** are vectors of the same size. The **x** values correspond to the horizontal axis, and the **y** values are represented on the vertical axis.

Several pairs of vectors (for the horizontal and vertical axes) can be provided to **plot**. MATLAB will plot the values given by the pairs of vectors on the same set of axes in the Figure window. If just one vector is provided as an argument, then the function plots the values in the vector against the index $1 \ldots n$, where n is the length of the vector. For example, the following command plots two curves

```
plot(x1,y1,x2,y2)
```

The first curve plots **y1** against **x1**, and the second shows **y2** versus the values in **x2**.

Many arguments can be used with the **plot** function, and these give the user a lot of control over the appearance of the graph. Most of them require the use of MATLAB's Handle Graphics® system, which is beyond the scope of this book. The interested reader is referred to Marchand and Holland [2002] for more details on Handle Graphics. However, we do include a discussion of how you can change the properties of a graph using the GUI plot editing tool in a later section of this chapter.

There is a lot of useful information in the MATLAB documentation; type **help plot** at the command line to see what can be done with **plot**. We present some of the basic options here. The default line style in MATLAB is a solid line, but there are other options as listed in Table 2.1. The first column in the table contains the notation for the line style that is used in the **plot** function. For example, one would use the notation

```
plot(x,y,':')
```

to create a dotted line with the default marker style and color. Note that the specification of the line style is given within single quotes (denoting a string) and is placed immediately after the vectors containing the observations to be graphed with the line style.

We can also specify different colors and marker (or point) styles within the single quotes. The predefined colors are given in Table 2.2. There are thirteen marker styles, and they are listed in Table 2.3. They include circles, asterisks, stars, x-marks, diamonds, squares, triangles, and more.

The following is a brief list of common plotting tasks:

- Solid green line with no markers or plotting with just points:

```
plot(x,y,'g'),  plot(x,y,'.')
```

- Dashed blue line with points shown as an asterisk:

```
plot(x,y,'b--*')
```

- Two lines with different colors and line styles (see Figure 2.2):

```
plot(x,y,'r:',x2,y2,'k.-o')
```

TABLE 2.1

Line Styles for Plots

Notation	Line Type
–	solid line
:	dotted line
-.	dash-dot line
--	dashed line

TABLE 2.2

Line Colors for Plots

Notation	Color
b	blue
g	green
r	red
c	cyan
m	magenta
y	yellow
k	black
w	white

TABLE 2.3

Marker Styles for Plots

Notation	Marker Style
.	point
o	circle
x	x-mark
+	plus
*	star
s	square
d	diamond
v	down triangle
^	up triangle
<	left triangle
>	right triangle
p	pentagram
h	hexagram

It is always good practice to include labels for all axes and a title for the plot. You can add these to your plot using the functions **xlabel**, **ylabel**, and **title**. The basic input argument for these functions is a text string:

```
xlabel('text'),ylabel('text'),title('text')
```

In the bulleted list above, we showed how to plot two sets of **x** and **y** pairs, and this can be extended to plot any number of lines on one plot. There is another mechanism to plot multiple lines that can be useful in loops and other situations. The command to use is **hold on**, which tells MATLAB to apply subsequent graphing commands to the current plot. To unfreeze the current plot, use the command **hold off**.

We sometimes want to plot different lines (or any other type of graph) in a Figure window, but we want each one on their own set of axes. See Figure 4.3 for an example. We can do this through the use of the **subplot** function. This function creates a matrix of plots (or sets of axes) in a Figure window, and the basic syntax is

```
subplot(m, n, p)
```

This produces **m** rows and **n** columns of axes (or plots) in one Figure window. The third argument denotes what set of axes (or plot) is active. Any plotting commands after this function call are applied to the **p**-th plot. The axes are numbered from top to bottom and left to right.

The following example shows how to create two plots that are side by side:

```
% Create the left-most plot—the first one.
subplot(1,2,1)
plot(x,y)

% Create the right-most plot—the second one.
subplot(1,2,2)
plot(x2,y2)
```

The first two input arguments for **subplot** tell MATLAB that this will be a matrix of 1×2 plots. In other words, we want to divide the plotting space into a matrix of plots with one row and two columns. We plot **x** and **y** in the first column with the first call to **subplot**, and we plot **x2** and **y2** in the second call to **subplot**.

We can access a previous plot by invoking the subplot again and specifying the desired plot in the third argument. For example, we could add a title to the first plot as follows

```
% Add a title to the first plot.
subplot(1,2,1)
title('My Plot')
```

2.1.2 Plotting 3-D Data

We can plot three variables in MATLAB using the **plot3** function, which works similarly to **plot**. In this case, we have to specify three arguments that correspond to the three axes. The basic syntax is

$$\texttt{plot3(x,y,z)}$$

where **x**, **y**, and **z** are vectors with the same number of elements. The function **plot3** will graph a line in 3-D through the points with coordinates given by the elements of the vectors. There is a **zlabel** function that can be used to add an axes label to the third dimension. See Figure 2.5 for an example.

In many cases, one of our variables is a response or dependent variable, and the other two are predictors or independent variables. Notationally, this situation is given by the relationship

$$z = f(x, y).$$

The z values define a surface given by points above the x–y plane.

We can plot this type of relationship in MATLAB using the **mesh** or **surf** functions. Straight lines are used to connect adjacent points on the surface. The **mesh** function shows the surface as a wireframe with colored lines, where the color is proportional to the height of the surface. The **surf** function fills in the surface facets with color.

The **surf** and **mesh** functions require three matrix arguments, as shown here

$$\texttt{surf(X,Y,Z), mesh(X,Y,Z)}$$

The **X** and **Y** matrices contain repeated rows and columns corresponding to the domain of the function. If you do not have these already, then you can generate the matrices using a function called **meshgrid**. This function takes two vectors **x** and **y** that specify the domains, and it creates the matrices for constructing the surface. The **x** vector is copied as rows of the **X** matrix, and the vector **y** is copied as columns of the **Y** matrix.

As stated previously, the color is mapped to the height of the surface using the default color map. The definition of the color map can be changed using the function **colormap**. See the **help** on **graph3d** for a list of built-in color maps. You can also add a bar to your graph that illustrates the scale associated with the colors by using the command **colorbar**.

MATLAB conveys 3-D surfaces on a 2-D screen, but this is just one view of the surface. Sometimes interesting structures are hidden from view, but we can change the view interactively via a toolbar button in the Figure window, as shown in Figure 2.1. We can also use the **view** function on the command line, as shown here

$$\texttt{view(azimuth, elevation)}$$

The first argument **azimuth** defines the horizontal rotation. The second argument **elevation** corresponds to the vertical rotation. Both of these are given in degrees. For example, we are looking directly overhead (2-D view) if the azimuth is equal to zero, and the elevation is given by 90 degrees.

Sometimes, it is easier to rotate the surface interactively to find a good view. We can use the ROTATE **3D** button in the Figure window, as shown in Figure 2.1. When the button is pushed, the cursor changes to a curved arrow. At this point, you can click in the plot area and rotate it while holding the left mouse button. The current azimuth and elevation is given in the lower left corner, as you rotate the axes.

FIGURE 2.1
Click on the ROTATE **3D** *button to rotate 3-D plots. The current azimuth and elevation is indicated in the lower left corner as the axes rotate.*

2.1.3 Examples

The set of examples in this section illustrate the use of the **plot** and **surf** functions. We will generate the coordinates to graph using the probability density function of a univariate and bivariate normal distribution. These distributions and how to work with them in MATLAB will be discussed in Chapter 4, but it is important to note at this time that the functions used to generate the normal distributions are part of the Statistics Toolbox.

The MATLAB code given here shows how to get the **x** and **y** vectors for two univariate normal distributions. The first one is a standard normal, and the second has a mean of zero and a standard deviation of two.

```
% To illustrate how to plot a curve, first
% generate the values for a normal distribution.
% This creates a standard normal probability
% distribution object.
stdnpd = makedist('Normal');

% Now, create one with different parameters.
npd = makedist('Normal','mu',0,'sigma',2);

% Define a vector for the domain.
x = -4:.01:4;
```

```
% Get the y values for both distributions.
y1 = pdf(stdnpd, x);
y2 = pdf(npd, x);
```

Next, we plot the vectors on the same set of axes, as shown here:

```
% Now, plot the curves on the same set of axes.
plot(x,y1,x,y2,'-.')
xlabel('X')
ylabel('PDF')
title('Plots of Normal Distributions')
legend('mu = 0, sigma = 1','mu = 0, sigma = 2')
```

The result is shown in Figure 2.2.

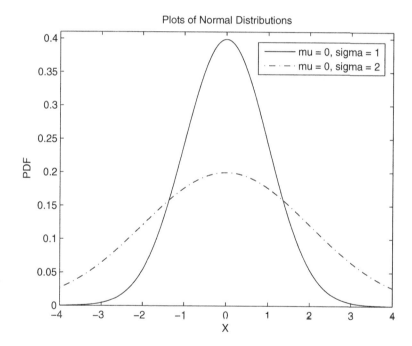

FIGURE 2.2

This is an example of plotting two curves on one set of axes. One is shown as a solid line, and the other is a dash-dot line. The legend was added using the **legend** *function.*

We now use the bivariate normal distribution to demonstrate how to create a surface plot. First, we have to set up the parameters of the distribution. This is for two dimensions. So, we need two values to specify the mean and a 2×2 symmetric matrix for the covariance. We will center the distribution at the origin, and we will use the identity matrix for the covariance.

```
% Get the vector for the mean.
mu = [0 0];

% Get the covariance matrix.
sigma = eye(2);
```

Next, we have to generate the (x, y) points for evaluating the distribution. The function for calculating the probability density function of a multivariate normal distribution requires a matrix, with each row containing an (x, y) coordinate. So, we first use **meshgrid** to get the **X** and **Y** domains and the notation **(:)** to convert each of these matrices to a vector. We then use **reshape** function to convert the results back to a matrix of the correct size for plotting using **mesh**.

```
% Obtain the (x,y) pairs for the domain.
x = -3:.2:3; y = -3:.2:3;
[X,Y] = meshgrid(x,y);

% Evaluate the multivariate normal at
% the coordinates.
Z = mvnpdf([X(:), Y(:)],mu,sigma);

% Reshape to a matrix.
Z = reshape(Z,length(x),length(y));
```

We now have what we need to create a surface plot for the given bivariate normal. The surface plot is shown in Figure 2.3.

```
% Now, create the surface plot and add labels.
surf(X,Y,Z);
xlabel('X'), ylabel('Y'), zlabel('PDF')
title('Multivariate Normal Distribution')
```

2.2 Scatter Plots

The scatter plot is one of the main tools a statistician should use before doing any analysis of the data or modeling (see Chapter 6). We describe 2-D and 3-D scatter plots in this section. We then discuss the scatter plot matrix for data with three or more dimensions. The Statistics Toolbox has special functions (**gscatter** and **gplotmatrix**) for grouped scatter plots that allow one to show groups in the data using different plotting symbols. Those functions will be described in Chapter 7.

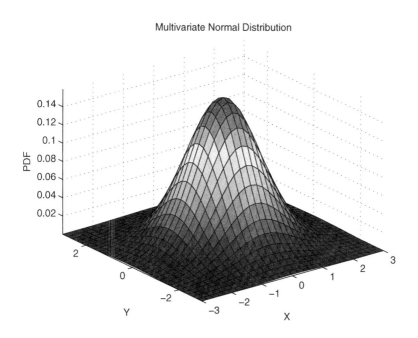

FIGURE 2.3
This is a surface plot of a bivariate normal distribution. The **surf** *function was used to create the plot.*

2.2.1 Basic 2-D and 3-D Scatter Plots

A *scatter plot* is just a plot of one variable against another, where (x, y) pairs are plotted as points. The points are not connected by lines. This type of plot can be used to explore the distribution of multivariate data, to assess the relationship between two variables, or to look for groups and other structure.

We can easily use the functions **plot** and **plot3** to create 2-D and 3-D scatter plots. We just specify the desired marker or symbol, as shown here:

$$\texttt{plot(x,y,'o'), plot3(x,y,z,'*')}$$

The call to **plot** shows the markers as open circles, and the call to **plot3** uses the asterisk as the plotting symbol.

The **plot** and **plot3** functions are best for the case where you are using only one or two marker styles and/or symbol colors. The base MATLAB software has two special functions for scatter plots that provide more control over the symbols used in the plot. These are called **scatter** and **scatter3**. The basic syntax is

$$\texttt{scatter(x,y,s,c), scatter3(x,y,z,s,c)}$$

The first two (or three for **scatter3**) arguments represent the coordinates for the points. The optional arguments **s** (marker size) and **c** (marker color) allow one to control the appearance of the symbols. These inputs can be a single value, in which case, they are applied to all markers. Alternatively, one could assign a color and/or size to each point.

The default symbol is an open circle. An additional input argument specifying the marker (see **help** on **plot** for a list) can be used to get a different plotting symbol. If we assign a different size to each of the open circles, possibly corresponding to some other variable, then this is known as a *bubble plot*.

2.2.2 Scatter Plot Matrix

In today's world, we often have data with many variables or dimensions. In this case, we can use the *scatter plot matrix* to look at all 2-D scatter plots, which gives us an idea of the pair-wise relationships or distributions in the data. A scatter plot matrix is just what the name implies—a matrix of plots, where each one is a 2-D scatter plot.

So, this is simply an example of what we saw previously with the **subplot** function. Fortunately, MATLAB provides a function called **plotmatrix** that takes care of the plotting commands for us, and we do not have to worry about multiple uses of **subplot**. The syntax for **plotmatrix** is:

$$\textbf{plotmatrix(X), plotmatrix(X,Y)}$$

The first of these plots the columns of **X** as scatter plots, with histograms of the columns on the diagonal. This is the version used most often by statisticians. The second plots the columns of **Y** against the columns of **X**. This is covered in more detail in Chapter 7.

2.2.3 Examples

We will use the **UStemps** and **iris** data sets to show how to construct scatter plots in MATLAB. First, we load the data into the workspace.

```
% Load the UStemps and iris data.
load UStemps
load iris
```

The next set of commands will construct a 2-D scatter plot. We changed the axes limits using the **axis** function so all points are easier to see. The plot is shown in Figure 2.4, and we see that there appears to be a linear relationship between temperature and latitude.

```
% Construct a 2-D scatter plot with plot,
% using temperature and latitude.
plot(Lat, JanTemp, '*')
```

```
% Adjust the axes to add white space.
axis([24 50 -2 70])

% Add labels.
xlabel('Latitude')
ylabel('Temperature (degs)')
title('Average January Temperature - US Cities')
```

The next example shows how to construct a 3-D scatter plot using **plot3**. We add grid lines and a box to help visualize the three dimensions. You can rotate the axes using the toolbar button, as described in the previous section. The scatter plot is shown in Figure 2.5.

```
% Construct a 3-D scatter plot using plot3.
plot3(Long, Lat, JanTemp, 'o')

% Add a box and grid lines to the plot.
box on
grid on

% Add labels.
xlabel('Longitude')
ylabel('Latitude')
zlabel('Temperature (degs)')
title('Average January Temperature - US Cities')
```

Our final example is a scatter plot matrix of the **iris** data. Recall that this data set has four variables. So, we will get a 4×4 matrix of scatter plots. See Figure 2.6 for the results.

```
% This produces a scatter plot matrix.
% We first need to put the data into one matrix.
irisAll = [setosa; versicolor; virginica];
plotmatrix(irisAll)
title('Iris Data - Setosa, Versicolor, Virginica')
```

2.3 GUIs for Graphics

MATLAB has several graphical user interfaces (GUIs) to make plotting data easier. We describe the main tools in this section, including simple edits using menu options in the Figure window, the plotting tools interface, and the PLOTS tab on the desktop ribbon.

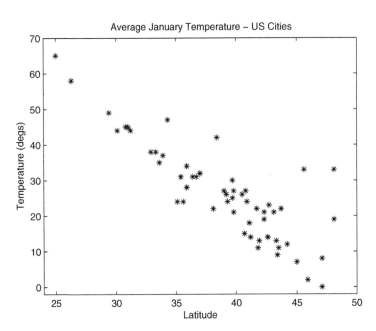

FIGURE 2.4
This is a 2-D scatter plot of the temperatures of US cities versus the latitude. Note that there appears to be a linear relationship with a negative slope.

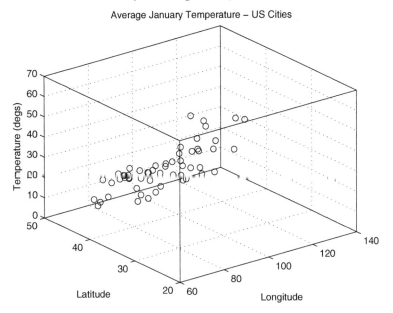

FIGURE 2.5
This shows a 3-D scatter plot of the temperatures of US cities versus longitude and latitude. The axes can be rotated interactively using the **ROTATE 3D** *button.*

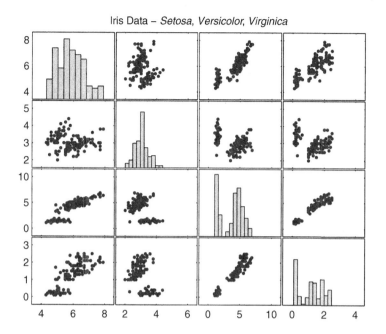

FIGURE 2.6

This is a scatter plot matrix for the iris data. It includes observations for all three species of iris. There does appear to be some evidence of groups in the data, possibly corresponding to the species of iris. A histogram for each of the variables (sepal length and width, petal length and width) and across all species is given along the diagonal.

<u>TIP</u>

You can save, print, or export your plot using options in the FILE menu of the Figure window. Use the EDIT menu in the Figure window to copy the plot for pasting into documents.

2.3.1 Simple Plot Editing

We often want to add simple graphics objects to our plot, such as labels, titles, arrows, rectangles, circles, and more. We can do most of these tasks via the command line, but it can be frustrating trying to get them to appear just the way we want them to. We can use the INSERT menu on the Figure window, as shown in Figure 2.7, to help with these tasks. Just select the desired option, and interactively add the object to the plot.

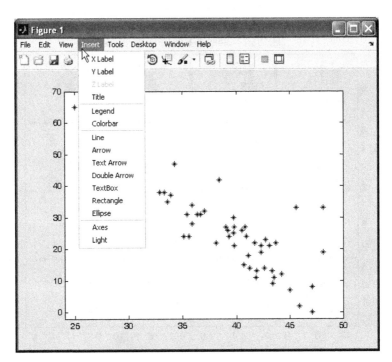

FIGURE 2.7
Select the INSERT menu on the Figure window to add objects to your plot.

2.3.2 Plotting Tools Interface

Perhaps the most comprehensive GUIs for working with graphics are the plotting tools. This is an interactive set of tools that work similarly to the main MATLAB desktop environment. In other words, they can be part of an expanded Figure window, or they can be undocked by clicking the downward arrow in the upper right corner of the tool. Look at **help** on **plottools** for more information and examples of using these tools.

The plotting tool GUIs consist of three panels or editors, as listed here:

- **Property Editor:** This provides access to some of the properties of the graphics objects in a figure. This includes the Figure window, the axes, line objects, and text. The editor can be started by using the command **propertyeditor**. It can also be opened via the TOOLS > EDIT PLOT menu item in a Figure window. Double-click on a highlighted graphics object to open the editor.

- **Figure Palette:** This tool allows the user to add and position axes, plot variables from the workspace, and annotate the graph. The command **figurepalette** will open the editor.

- **Plot Browser**: The browser is used to add data to the plot and to control the visibility of existing objects, such as the axes. This can be opened using the command `plotbrowser`.

The plotting tools can be opened in different ways, and we specified some of them in the list given above. One option is to use the command `plottools`. This will add panels to an existing Figure window, or it will open a new Figure window and the tools, if one is not open already. The most recently used plotting tools are opened for viewing.

One could also click on the highlighted toolbar button shown in Figure 2.8. The button on the right shows the plotting tools, and the one on the left closes them. Finally, the tools can be accessed using the VIEW menu on the Figure window. Clicking on the desired tool will toggle it on or off.

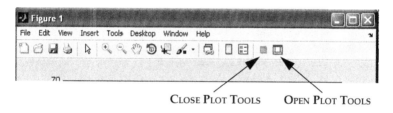

CLOSE PLOT TOOLS   OPEN PLOT TOOLS

FIGURE 2.8

This shows the toolbar buttons for a Figure window. The buttons on the right will open and close the interactive plotting tools.

An example of a Figure window with the plotting tools open is given in Figure 2.9. Only the Property Editor is opened, because that was the last tool we used. Select other editors using the VIEW menu.

2.3.3 PLOTS Tab

Another GUI option to create plots is available via the **PLOTS** tab on the main MATLAB ribbon interface. The Workspace browser has to be open because this is used to select variables for plotting. Click on the desired variables in the browser, while holding the **CTRL** key. The variables will appear in the left section of the **PLOTS** tab.

Next, select the type of plot by clicking the corresponding picture. There is a downward arrow button that provides access to a complete gallery of plots. It will show only those plots that are appropriate for the types of variables selected for plotting. See Figure 2.10 for an example.

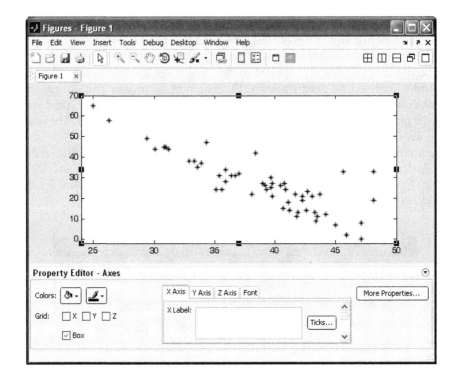

FIGURE 2.9

Here is an example of a Figure window with the potting tools open. Only the Property Editor is viewed because this is the last one that was used. Select other editors using the **VIEW** *menu.*

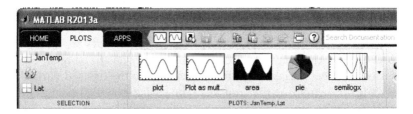

FIGURE 2.10

This is a portion of the **PLOTS** *ribbon in the main MATLAB desktop environment. Click the* **PLOTS** *tab to access it. Variables for plotting are selected in the Workspace browser, and they appear in the left section of the ribbon. Click on one of the icons in the right section of the* **PLOTS** *ribbon to create the plot. More options are available using the downward arrow button in the plots section.*

2.4 Summary and Further Reading

MATLAB has more graphing functions as part of the base software. The main ones were described in this chapter, and we provide an expanded list in the following table. Use the **help *functionname*** at the command line for information on how to use them and to learn about related functions. Table 2.5 has a list of auxiliary functions for enhancing your plots.

TABLE 2.4

List of Plotting Functions in Base MATLAB®

area	Plot curve and fill in the area
bar, bar3	2-D and 3-D bar plots
contour, contour3	Display isolines of a surface
errorbar	Plot error bars with curve
hist	Histogram
image	Plot an image
pie, pie3	Pie charts
plot, plot3	2-D and 3-D lines
plotmatrix	Matrix of scatter plots
scatter, scatter3	2-D and 3-D scatter plots
stem, stem3	Stem plot for discrete data

TABLE 2.5

List of Auxiliary Plotting Functions in Base MATLAB®

axis	Change axes scales and appearance
box	Draw a box around the axes
grid	Add grid lines at the tick marks
gtext	Add text interactively
hidden	Remove hidden lines in **mesh** plots
hold	Hold the current axes
legend	Insert a legend
plotedit	Tools for annotation and editing
rotate	Rotate using given angles
subplot	Include multiple axes in figure window
text	Insert text at locations
title	Put a title on the plot
xlabel, ylabel, zlabel	Label the axes
view	Specify the view for a 3-D plot
zoom	Zoom in and out of the plot

We already recommended the Marchand and Holland book [2002] for more information on Handle Graphics. This text also has some useful tips and ideas for creating plots and building GUIs. For more background on how to visualize data in MATLAB, please see the book *Exploratory Data Analysis with MATLAB* by Martinez et al. [2010]. Finally, you should always consult the MATLAB documentation and help files for examples. For example, there is a section called Graphics in the MATLAB documentation center. Recall that you get to the documentation by clicking the HELP button in the RESOURCES section of the HOME ribbon and selecting the MATLAB link.

We now provide some references to books that describe scientific and statistical visualization, in general. One of the earliest ones in this area is called the *Semiology of Graphics: Diagrams, Networks, Maps* [Bertin, 1983]. This book discusses rules and properties of graphics. For examples of graphical mistakes, we recommend the book by Wainer [1997]. Wainer also published a book called *Graphic Discovery: A Trout in the Milk and Other Visual Adventures* [2004] that details some of the history of graphical displays in a very thought provoking and entertaining way. The book *Visualizing Data* [Cleveland, 1993] includes descriptions of visualization tools, the relationship of visualization to classical statistical methods, and some of the cognitive aspects of data visualization and perception. Another excellent resource on graphics for data analysis is Chambers et al. [1983]. Finally, we highly recommend Naomi Robbins' [2005] book called *Creating More Effective Graphs*. This text provides a wonderful introduction on ways to convey data correctly and effectively.

There is a *Graphics* section in the online documentation for base MATLAB. Recall that you can access this documentation via the HELP button on the RESOURCES tab. The Statistics Toolbox documentation has a chapter called *Exploratory Data Analysis,* under which is a section on *Statistical Visualization*. This has details about univariate and multivariate plots.

Chapter 3

Descriptive Statistics

I abhor averages. I like the individual case. A man may have six meals one day and none the next, making an average of three meals per day, but that is not a good way to live.
Louis Brandeis (1856–1941)

Now that we know how to import our data (see Chapter 1), the next step in statistical analysis is to understand various aspects of our data better. For example, it would be helpful to have an idea of how the data are distributed before conducting hypothesis tests or constructing models. In this chapter, we illustrate how to estimate some statistics that describe the location and scale of our data, along with ways to visually understand how the data are distributed.

However, it is important to define two concepts before we delve into the topic of descriptive statistics. These are ***populations*** and ***samples***. Both of these are a collection of measurements of some attribute or variable that we are interested in studying. A population includes measurements for *all* instances (past, present, or future), while a sample contains only *some* of them. A sample is meant to be representative of a population, and we often use it to describe the population and to make inferences about it.

3.1 Measures of Location

Measures of location are also known as ***measures of central tendency***. We can think of these statistics as an estimate of a typical value for the data. In other words, if our data included measures of the heights (an attribute or variable) of the students in a class, then a measure of central tendency would indicate a typical height for the students. We will show how you can use MATLAB® to calculate measures of central tendency for quantitative or numeric variables.

3.1.1 Means, Medians, and Modes

The base MATLAB package has functions for calculating three measures of central tendency—mean, median, and mode. The first two are suitable for quantitative data, while the mode can be used for either type.

There are several types of means, and we will discuss the most common one. This is the arithmetic mean, which is sometimes called the **sample mean** or the **average**. It is calculated by adding up all of the observed values of our attribute and dividing by the total number of observations. The formula for this type of mean is given here:

$$\bar{x} = \frac{1}{n}\sum_{i=1}^{n} x_i,$$

where the letter x_i denotes the i-th observed value of the attribute, n is the number of data points we have in our data set, and $\bar{x}$ represents the sample mean.

TIP

There are two other types of means, besides the sample (or arithmetic) mean. These are called the *geometric mean* and the *harmonic mean*. The Statistics Toolbox has functions to calculate these quantities. The functions are called **geomean** and **harmmean**.

The sample mean can be affected by extreme values—either very large or very small. Say you are averaging your test scores in a class, and each test is equally weighted for the final grade. If you get a really bad grade (for example a zero) on a test, then what does that do to your final grade? That one bad grade will pull your average down.

The same thing can happen with real data. We might have an attribute that comes from a skewed distribution (more on this later in the chapter). Or, we might have some data entry errors. In these cases, we might want to use a different measure of location, such as the median or the trimmed mean.

The **sample median** is essentially the middle point of the data. In other words, it is a number that divides the data into two equal sub-groups. To find the median, we first order the data from smallest to largest. If we have an odd number of observations, then the sample median is the middle data point. If we have an even number of observations, then the sample median is the average of the two middle points.

The **trimmed mean** is calculated using the same process we used for the sample mean. However, we first remove a small percentage ($k\%$) of the highest and lowest values from our data set. This has the effect of removing potential extreme values that might adversely affect our estimate of a typical value for our data.

The median and the trimmed mean are both resistant to extreme values, but the median is more resistant than a typical trimmed mean. We would need to have 50% of the observations as extremely large (or small) before the median would be pulled in that direction. On the other hand, if more than $k\%$ of our observed values are extreme, then the trimmed mean would be affected by them.

The *mode* is defined as the observation that occurs with the highest frequency in the data set. Unlike the mean and the median, the mode can be used to give us a typical value in our data set when the data are qualitative (non-numeric) in nature, as well as quantitative. There is currently no easy way in MATLAB to get the mode of qualitative values.

3.1.2 Examples

We will use the density of the Earth data set to demonstrate MATLAB code you can use to estimate the measures of location that we just described. Recall that we saved the Earth density data as a variable called **earth** in a **.mat** file using the methods discussed in Chapter 1. We first import the data into our workspace and then calculate the sample mean.

```
% We will first load the variable.
load earth
% Now, we find the mean.
xbar = mean(earth)

xbar =

    5.4197
```

Thus, a typical measure of Earth's density is 5.4197 (times the density of water).

Now, let's see what happens when we find the median. We can find this in MATLAB, as follows:

```
% Find the median density of the Earth.
med = median(earth)

% This is the value we see in the command window.
med =

    5.4600
```

So, we have a slightly higher measure of location using the median. We will see in a later section when we visualize the data why this is so. The difference between the mean and the median gives us some evidence that there might be some extreme values in the data.

The MATLAB functions **mean** and **median** are included in the base MATLAB package, but the function to calculate the trimmed mean is part of the Statistics Toolbox. The basic syntax for this function is

```
trimmean(x, percent)
```

This will remove the $K = n \times (k/100)/2$ highest and lowest values. We use 20% in the following calculation for the trimmed mean. This removes the top 10% and smallest 10% of the values in the data.

```
% Find the trimmed mean density of the Earth.
% We will use 20% for our trimming.
xbar_trim = trimmean(earth, 20)


xbar_trim =

    5.4617
```

We see that this estimate of central tendency is closer to the median, as we would expect.

We can also find the mode of the **earth** data using the **mode** function. Here is the code.

```
% Now, find the mode.
mod_earth = mode(earth)


mod_earth =

    5.2900
```

This gives a value of 5.29 times the density of water, which is quite a bit less than the sample mean. There could be more than one mode in the data, if more than one value occurs with the same highest frequency. In this case, MATLAB will return the smallest of the modes.

Thus, it would be a good idea to look for other modes, and we can use the MATLAB Variable Editor to do this (see Figure 3.1). First, we sort the data using the toolbar button on the ribbon, and we see that there are two modes: 5.29 and 5.34. The second mode is a little closer to our sample mean.

3.2 Measures of Dispersion

We just learned that we can use something like the sample mean to give us a typical value for a data set. However, it is likely that no value in the data set will exactly equal the mean because there will be some variation in the data.

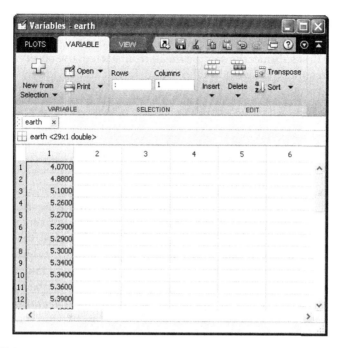

FIGURE 3.1

This is a screenshot of the MATLAB Variable Editor, showing the sorted **earth** *variable. We sorted the data using the* **Sort** *button on the toolbar ribbon. We can then see the two modes at 5.29 (rows 6 and 7) and 5.34 (rows 9 and 10).*

This section will illustrate some of the functions MATLAB provides for finding *measures of variation* or *dispersion* for the data.

3.2.1 Range

Some very simple descriptive statistics that give us some idea of the spread of our data are the maximum value, the minimum value, and the range. The *range* is calculated by subtracting the minimum value of our observed data points from the maximum value, as shown here:

$$range = x_{max} - x_{min}.$$

However, this is not a very efficient or informative measure, because it excludes a lot of data from the calculation.

3.2.2 Variance and Standard Deviation

A better measure of data dispersion is the sample *variance*. To calculate this statistic, we first subtract the sample mean from each of the observations and square the result. Next, we add all of these values and divide by $n - 1$. The formula is given by:

$$\hat{\sigma}^2 = \frac{1}{n-1} \sum_{i=1}^{n} (x - \bar{x})^2.$$

This is the average (approximately) of the squared deviations from the sample mean. Thus, it measures the dispersion around the mean.

The variance gives us the spread in squared units of our measured attribute. For example, if the observations are heights in inches, then the units of the variance would be squared inches. This is not a very helpful quantity when associated with the average height in inches. We can take the square root of the variance to get a measure of spread that would be in the same units as the observed variable. This is called the **standard deviation**.

It should not be surprising, given the definition, that the variance (or standard deviation) is usually reported as a measure of dispersion with the mean. There is also a measure of dispersion that is associated with the median, and that is called the interquartile range. We discuss that statistic in a later section.

TIP

The Statistics Toolbox has a function called mad that computes the mean absolute deviations, which is another way to measure variability in the data.

3.2.3 Covariance and Correlation

So far, we have been talking about a single attribute or variable. However, we often have a data set containing several variables, such as the Fisher's iris data set described in Chapter 1. In this case, we could calculate the sample covariance or the correlation coefficient, which measure the joint variation of two variables.

The sample *covariance* between two variables X and Y is given by

$$cov(X, Y) = \frac{1}{n-1} \sum_{i=1}^{n} (x_i - \bar{x})(y_i - \bar{y}).$$

We see from this equation that the covariance depends on the product of the relationship of the X and Y variables with their respective means. If large values of X tend to be associated with large values of Y, then the covariance will be positive. The same thing holds if small values of X are associated with small values of Y. This happens, because the products in the summation will be mostly positive. We could have the opposite situation, where small values of X tend to correspond with large values of Y, or vice versa. In this case, the products are predominantly negative, and the summation will likely be negative.

The sign of the covariance measures the direction of the linear relationship between two variables. The magnitude of the covariance measures the strength of the relationship, but it is difficult to interpret this value. This is partly because it depends on a function of the variable units. It would be desirable to have a statistic that is unitless, which we can get by dividing the covariance by the standard deviations for X and Y.

When we do this division (or normalization), we get the sample *correlation coefficient*, which is given by

$$corr(X, Y) = \frac{cov(X, Y)}{\sqrt{\hat{\sigma}_X^2}\sqrt{\hat{\sigma}_Y^2}}.$$

The correlation coefficient has some nice properties. It can only take on a value between -1 and 1. A correlation coefficient close to the extreme values indicates a strong linear relationship between the two variables, and it will equal -1 or 1 when X and Y are related by a straight line. The sign of the correlation coefficient indicates the direction of the relationship—either positive or negative slope.

3.2.4 Examples

We now use MATLAB to calculate these descriptive statistics for the **earth** data. All of the functions shown here are part of the base MATLAB software package.

```
% First find the minimum, maximum and range
% of the earth data.
minearth = min(earth)

minearth =
     4.0700

maxearth = max(earth)

maxearth =
     5.8600
```

```
rngearth = range(earth)

rngearth =
    1.7900
```

Just as a check, we could find the range manually by subtracting **minearth** from **maxearth**, and we get the same answer of 1.79.

```
% We can find the range, as follows
maxearth - minearth

ans =

    1.7900
```

Next, we calculate the variance and the standard deviation of the data using the MATLAB functions **var** and **std**.

```
% Next, we find the variance and the
% standard deviation.
vearth = var(earth)

vearth =

    0.1148

searth = std(earth)

searth =

    0.3389
```

We could also find the standard deviation by taking the square root of the variance.

```
% We can also find the standard deviation
% by taking the square root of the variance.
sqrt(var(earth))

% This is displayed at the command line.
ans =

    0.3389
```

Recall that the covariance and correlation statistics measure the association between two variables. So, we need to use a different data set to demonstrate how these can be calculated using MATLAB. The Fisher's **iris.mat** data file has three objects, each of which has four variables, and we will use one of these objects to show how to get these statistics.

The MATLAB functions **cov** and **corrcoef** will produce matrices, where the *i,j*-th element of the matrix contains the appropriate statistic for the *i*-th and *j*-th variables. For example, the element in the first row ($i = 1$) and the second column ($j = 2$) of the covariance matrix is the covariance between the first and second variables. The diagonal elements of the covariance matrix correspond to the variances of the individual variables, and the diagonal elements of the correlation coefficient matrix will be equal to 1.

```
% Now, let's find the covariance matrix of the
% setosa iris. First, we need to load it.
load iris

% Find the covariance matrix of the setosa object.
cvsetosa = cov(setosa)

cvsetosa =

    0.1242      0.0992      0.0164      0.0103
    0.0992      0.1437      0.0117      0.0093
    0.0164      0.0117      0.0302      0.0061
    0.0103      0.0093      0.0061      0.0111

% Find the correlation coefficients.
crsetosa = corrcoef(setosa)

crsetosa =

    1.0000      0.7425      0.2672      0.2781
    0.7425      1.0000      0.1777      0.2328
    0.2672      0.1777      1.0000      0.3316
    0.2781      0.2328      0.3316      1.0000
```

As stated before, the correlation coefficients are easier to interpret. We see that the magnitude of the statistics are all positive, indicating that the slope of the linear relationship is positive for all pairs of variables. We can also see that the relationship between the first two variables (sepal length and sepal width) is somewhat strong, because the correlation coefficient for these two variables is 0.7425, which is getting close to 1.

Also, note that the diagonal elements of the **crsetosa** are all equal to 1, as stated previously. We can use the **var** function to verify that the diagonal elements of the covariance matrix are equal to the variances of each variable.

```
% If the argument to the var function is a matrix,
% then it will return the variance of each column
% or variable.
```

```
var(setosa)

ans =

    0.1242      0.1437      0.0302      0.0111
```

As expected, these match the diagonal elements of the covariance matrix.

TIP

The Statistics Toolbox has a function called corr that has additional options and information to help in data analysis.

3.3 Describing the Distribution

The measures of central tendency and dispersion tell us something about the distribution of the data, but there are other aspects that can be informative, such as the shape of the distribution. This type of information is important when it comes to analyzing our data, because many of the methods rely on assumptions made about the distribution of the data. In this section, we will describe statistics based on quantiles, such as the quartiles, percentiles, and the interquartile range. We conclude with a discussion of the sample skewness.

3.3.1 Quantiles

The quantiles are important characteristics of a distribution, and they are used in many data analysis methods. For example, quantiles are the critical values used in hypothesis testing (see Chapter 5) and in the estimation of confidence intervals [Kotz and Johnson, 1986].

Essentially, the sample *quantiles* divide the data set into q equal-sized groups of adjacent values. To determine the adjacent values, we put the data in ascending order, which produces the order statistics denoted by

$$x_{(1)}, x_{(2)}, \ldots, x_{(n)}.$$

We have already talked about one quantile—the median—that divides the data into two groups, with half of the ordered data located above the median and half below it. This would correspond to $q = 2$.

Sometimes a sample quantile will correspond to one of the observations in our data set, but it is more likely to fall in between two order statistics. In this case, MATLAB uses linear interpolation to calculate the sample quantile. It is important to note that most statistical software systems do not use a common definition for the sample quantiles. To confuse things further, some packages use different definitions within their own software system, depending on how the quantile will be used (e.g., in a boxplot or a Q–Q plot) [Hyndman and Fan, 1996].

A common quantile of interest is the ***percentile***, which corresponds to $q = 100$. We need 99 values to divide the data into 100 equal-sized groups. We often see this statistic reported with test score results, where we are given what percentile the score falls in.

Other quantiles that are used a lot in statistical analysis are the ***quartiles***, with $q = 4$. There are three sample quartiles that divide the data into four equal-sized groups, with 25% of the data in each group. We will denote these sample quartiles as $\hat{Q}_{0.25}, \hat{Q}_{0.50}, \hat{Q}_{0.75}$. We already encountered $\hat{Q}_{0.50}$; this is the sample median.

The MATLAB Statistics Toolbox has two functions for finding quantiles. These are called **quantile** and **prctile**. Both of these functions can be used to find the quartiles, as we will show in the examples.

3.3.2 Interquartile Range

Like the standard deviation, the interquartile range quantifies the dispersion or spread in the data. It is typically reported when the median is given as the measure of central tendency, rather than the standard deviation. The sample *interquartile range* or IQR is defined as

$$\hat{Q}_{0.75} - \hat{Q}_{0.25}.$$

Thus, we see that this is the difference between the third and first quartiles or the range of the middle 50% of the data.

Like the median, the IQR is resistant to outliers or extreme values in the data set. For example, making any of the observations in the first or last sub-groups more extreme will not affect the values of the first and third quartiles or the IQR. We could think of this as trimming the upper 25% and lower 25% of the data and calculating the range. The MATLAB Statistics Toolbox has a function **iqr** that will return the IQR of a data set.

3.3.3 Skewness

Shape is an important property of a distribution, and we can describe the shape as being symmetric, skewed, flat, and more. The normal distribution (or a bell-shaped curve) is ***symmetric***, which means that the curve on both sides of the mean are mirror images of each other. A skewed distribution is ***asymmetric***, because it does not have this property.

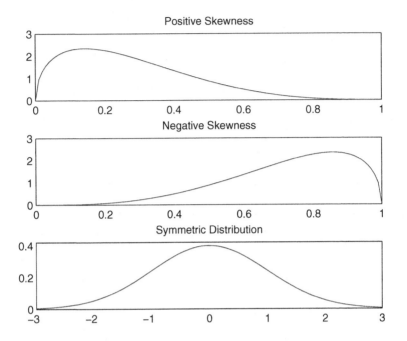

FIGURE 3.2
This figure shows three different distribution shapes. The first one corresponds to a distribution that has the long tail to the right and is positively skewed. The second distribution is one that has negative skewness or a long tail to the left. The third example shows a symmetric distribution. In this case, both sides of the curve will fall on top of each other, if we reflect the distribution around zero (the mean).

Skewness is a measure of how a distribution departs from symmetry. A distribution that is skewed has a long tail in one direction and usually has one mode (area of higher density) in the other direction. A distribution that is negatively skewed has a long tail to the left, and a distribution that is positively skewed has a long tail to the right. Figure 3.2 provides examples of distributions with different shapes.

The sample skewness is given by

$$\frac{\frac{1}{n}\sum (x_i - \bar{x})^3}{\left(\frac{1}{n}\sum (x_i - \bar{x})^2\right)^{3/2}} .$$

The skewness will be negative for distributions that are negatively skewed, and it will be positive for ones that are positively skewed. For symmetric distributions, the sample skewness will be close (or equal) to zero. The MATLAB Statistics Toolbox has a function called **skewness** that will calculate the sample skewness for a data set.

3.3.4 Examples

We continue to use the **earth** data to illustrate the descriptive measures we just discussed. We first use the **quantile** function to determine the sample quartiles for the **earth** data, as shown here. The syntax is

quantile(data,P)

where **P** is a vector of cumulative probability values between zero and one. We also re-calculate the median of the sample to show that this is the same as the second quartile.

```
% First, we find the quartiles.
quart = quantile(earth,[.25 .50 .75])

quart =

   5.2975    5.4600    5.6125

median(earth) % This gives the same result.

% This is what we get from the median function.
ans =

   5.4600
```

Next, we find the interquartile range using the **iqr** function.

```
% Next, find the IQR of the earth data.
iqrearth = iqr(earth)

iqrearth =

   0.3150
```

Thus, we can report robust measures of central tendency and dispersion using the median and the IQR. For example, the median density of the earth is 5.46 (as a multiple of the density of water), with an interquartile range of 0.315.

The **prctile** function works in the same way as **quantile**, but the inputs are specified as percentages or numbers between 0 and 100. Thus, we can also find the quartiles of the **earth** data as follows.

```
pearth = prctile(earth,[25 50 75])

pearth =

    5.2975     5.4600     5.6125
```

We describe ways to explore our data visually in the next section, and we will see that the **earth** data is skewed. We can verify that by calculating the sample skewness.

```
% Find the sample skewness of the earth data.
skearth = skewness(earth)

skearth =

   -2.2068
```

The result is negative, indicating that the distribution has a long tail to the left or that the data are more spread out to the left of the mean.

3.4 Visualizing the Data Distribution

So far in this chapter, we have been focusing on descriptive measures and summary statistics. It is always a good idea to also explore the data using visualization techniques. In this section, we will look at ways one can graph a single variable in order to understand how it is distributed. This includes histograms, Q–Q plots, and boxplots.

3.4.1 Histograms

One of the quickest and easiest ways to visualize the distribution of a single variable is to construct a histogram. A *histogram* uses vertical bars (or bins) to summarize the distribution of the data. The bars are adjacent to each other and do not have any space in between them.

There are two basic types of histograms. The first is called the *frequency histogram*, where the heights of the bars reflect the number of observations that fall into the interval defined by the bar. The second type is the *relative frequency histogram*, where the heights of the bars correspond to the number of observations in the interval divided by the total number of observations. These two histograms will convey the same shape for the distribution—only the scale for the vertical axis changes.

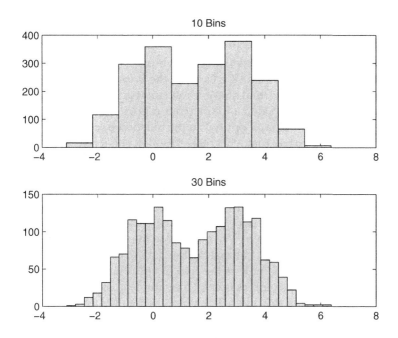

FIGURE 3.3
The histogram at the top shows the frequency distribution of a data set with 10 bins. The frequency histogram at the bottom represents the distribution of the same data set using 30 bins or bars. We can see the same general shape, but there is more structure visible in the bottom one.

The structure one can see in histograms depends on the number of bins or bars. The base MATLAB package has a function called **hist**, and it will create a histogram with a default value of 10 bins. Figure 3.3 shows an example of two frequency histograms for the same data set. The histogram at the top uses the default, and the one at the bottom has 30 bins. One can see additional structure (e.g., peaks and troughs) when we add more bins.

3.4.2 Probability Plots

There are two types of plots that will visually compare the distribution of a data set to a theoretical distribution. These include the probability plot and the Q–Q plot. Both of these plots follow the same general approach, where we plot the quantiles of one distribution against the quantiles of another as points. If the points fall roughly on a straight line, then the distributions are approximately the same.

To construct a *probability plot*, one must specify a theoretical distribution for comparison to the sample quantiles, which are just the ordered data. The

MATLAB Statistics Toolbox has a function called **probplot** that will compare the sample distribution to the normal, exponential, and other theoretical distributions. (See Chapter 4 for information on probability distributions.) When we conduct an analysis of our data, we often have to determine if our data are normally distributed. We can use the *normal probability plot* to compare our data to the normal distribution, but it is really just a special case of the probability plot. One can construct a normal probability plot in MATLAB by using the default options in the function **probplot**. There is a function called **normplot** that will create this plot, and it also adds a straight line that goes through the first and third quartiles. This helps the analyst assess the linearity of the points.

A Q–Q plot (or *quantile-quantile plot*) was first developed by Wilk and Gnanadesikan in 1968 [Wilk and Gnanadesikan, 1968]. It is a way to visually compare the distributions of two variables in a data set, and it is constructed similarly to the probability plot. We do not use the theoretical quantiles in a Q–Q plot because we are not comparing the data to a specified distribution. Rather, we would like to know if the two distributions that generated the data are the same. So, we plot (as points) the sample quantiles of one variable against the quantiles of the other. The points will fall roughly on a straight line if the data are distributed in the same manner. The **qqplot** in the MATLAB Statistics Toolbox will construct one of these plots.

3.4.3 Boxplots

Boxplots were developed by John Tukey [1977], and they have been used for many years. They offer an excellent visual summary of a data set. This summary shows the quartiles, the IQR, the maximum and minimum values, any potential outliers, and the distribution shape.

A vertical boxplot is comprised of a box with the length given by the IQR. Horizontal lines are placed at the quartiles, and vertical lines are added to create a box. A line is extended from the first quartile to the smallest adjacent value, and a similar line connects the third quartile to the largest adjacent value. *Adjacent values* are the most extreme observations in the data set that are not potential outliers. The adjacent values can be defined differently in statistical software packages, but are usually found by adding/subtracting a multiple of the IQR to the third/first quartiles. Any potential outliers are added as points to the boxplot. An illustration of a boxplot is shown in Figure 3.4.

A boxplot can be created to summarize the distribution of one variable or we could also use it to compare several variables by constructing side-by-side boxplots. We can also add a notch to the boxplot, which conveys the end points of a 95% confidence interval for the median of each variable. If these intervals do not overlap, then this is significant evidence that the medians are

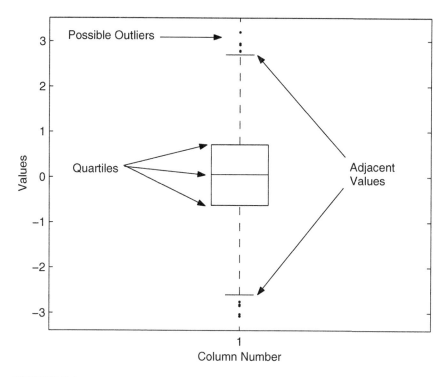

FIGURE 3.4

This is a schematic of a vertical boxplot with possible outliers. Boxplots can also be shown horizontally. We can see robust summaries of the data—the median, the IQR, and more. We can also determine if the distribution is skewed or symmetric [Martinez et al., 2010].

different. See Chapter 5 for information on hypothesis testing and confidence intervals.

3.4.4 Examples

We return to the **earth** data set to show how to construct a histogram using MATLAB. We use the **hist** function in base MATLAB to get a histogram with the default of 10 equally spaced bins. This is shown in Figure 3.5. Note that the distribution appears to be skewed to the left or negatively skewed. This confirms the sample skewness calculated in the previous section.

```
% Create a histogram of the earth data
% using 10 bins.
hist(earth)
title('Histogram of the Earth Density Data')
xlabel('Multiple of the Density of Water')
ylabel('Frequency')
```

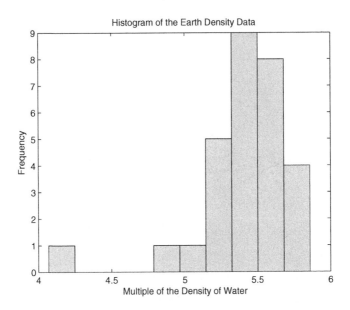

FIGURE 3.5
This is a histogram of the **earth** *data, using the default of 10 equally spaced bins. Note that the distribution of the data looks skewed to the left.*

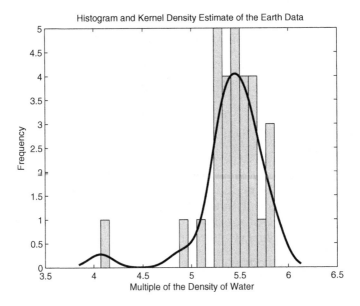

FIGURE 3.6
This is another histogram of the **earth** *data, where we use 20 bins and superimpose a smooth estimate of the density function.*

The MATLAB Statistics Toolbox has a function **histfit** that will create histograms, and it has more capability than **hist**. In particular, **histfit** will superimpose a smooth density function based on a specified distribution (such as the normal distribution) or it will estimate a function using the kernel density estimation approach [Scott, 1992]. We apply this to the **earth** data and show the results in Figure 3.6.

```
% Use the histfit function in the
% Statistics Toolbox.
% Request a kernel density estimate along with
% the histogram, and use more bins.
histfit(earth,20,'kernel')
title('Histogram and Density of the Earth Data')
xlabel('Multiple of the Density of Water')
ylabel('Frequency')
```

The **earth** data does not look normally distributed, but we will get a probability plot for the normal distribution to check. The following MATLAB code will display the probability plot shown in Figure 3.7.

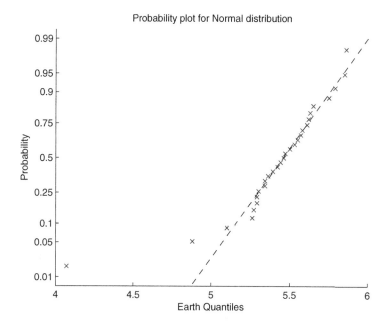

FIGURE 3.7

*This is a probability plot comparing the quantiles of the **earth** data to the quantiles of a theoretical normal distribution. The data do not appear to be normally distributed.*

```
% Get a probability plot comparing the
% earth data to a normal distribution.
probplot(earth)
xlabel('Earth Quantiles')
```

We see that it does not appear to be normally distributed because of the long tail. There also seems to be some extreme or outlying values in the lower part of the distribution. This matches the appearance of the distribution in the histograms.

We will use Fisher's iris data to demonstrate how to create Q–Q plots and boxplots. First, we create a Q–Q plot to determine if the distribution of the sepal length for *Virginica* and *Versicolor* are the same.

```
% Construct a Q-Q plot of the sepal length
% for Virginica and Versicolor in the iris data.
qqplot(virginica(:,1),versicolor(:,1))
title('Q-Q Plot of Sepal Length - Iris Data')
xlabel('Virginica Quantiles')
ylabel('Versicolor Quantiles')
```

The plot is shown in Figure 3.8, and we see that the points are close to the line. This provides some evidence that the distributions are approximately the same apart from any differences in location and scale.

The following MATLAB steps will construct side-by-side notched boxplots. As we discussed previously, this allows us to perform a visual hypothesis test for the difference in the medians.

```
% Create boxplots of sepal length
% for the iris data.
boxplot([setosa(:,1),virginica(:,1),...
    versicolor(:,1)],...
    'notch','on',...
    'labels',{'Setosa','Virginica','Versicolor'})
title('Notched Boxplots of Sepal Length')
```

The boxplots are shown in Figure 3.9. Note that the intervals represented by the notches do not overlap. Therefore, we have significant evidence that the median sepal length for the three species are different.

3.5 Summary and Further Reading

The functions for descriptive statistics and visualization that were described in this chapter are summarized below in Tables 3.1 and 3.2. We also include additional MATLAB functions that can be used to explore and describe data.

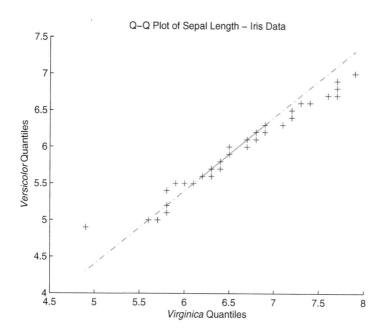

FIGURE 3.8

This Q–Q plot compares the distributions of the sepal length for Versicolor *and* Virginica *in Fisher's iris data. The points fall near the straight line, indicating that the data are distributed similarly apart from any differences in location and scale.*

As with every function, the reader should type **help functionname** for information on how to use it, what is produced, and any related functions.

There are many good books to help the reader understand basic statistics, including some that are available on the Internet. We list two of them here.

- *VassarStats:* **http://vassarstats.net/index.html**
- *Online Statistics Education: An Interactive Multimedia Course of Study:* **http://onlinestatbook.com/index.html**

The topic of descriptive statistics is typically included in most introductory statistics books. So, we do not provide specific references here.

However, it is worthwhile mentioning several books that cover some of the visualization methods described in this chapter. The seminal book on the use of visualization for exploratory data analysis was written by Tukey [1977]. Another excellent book is by Cleveland [1993]. For an in-depth discussion of visualization and exploratory data analysis using MATLAB, see Martinez and Martinez [2007] and Martinez, et al. [2010].

There is a section on *Descriptive Statistics* in the documentation for the base MATLAB software. You can access it by clicking the **Help** button in the **Resources** section of the **Home** ribbon and selecting the MATLAB link. The

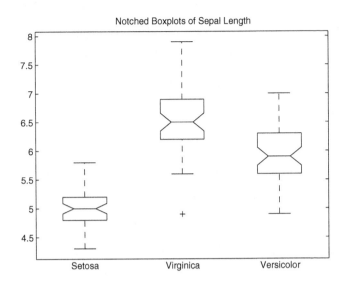

FIGURE 3.9

These are side-by-side notched boxplots, where the notches correspond to the end points of a 95% confidence interval for the median. The notched intervals do not overlap, indicating that the medians are significantly different. We also see a potential outlier for Virginica, as indicated by the plus symbol.

TIP

Base MATLAB has functions that will create a pie chart and a bar chart that can be used to examine the distribution. These are called `pie` and `bar`.

---

relevant information is under the *Mathematics*, followed by the *Statistics and Random Numbers* section.

The documentation for the Statistics Toolbox is accessed by clicking on the HELP button on the HOME ribbon and then clicking on the Statistics Toolbox link. The main section you want for techniques discussed in this chapter is called *Exploratory Data Analysis*. There are two relevant sections under this heading—*Descriptive Statistics* and *Statistical Visualization*.

TABLE 3.1

List of Functions for Describing Data in Base MATLAB®

Statistics	
`corrcoef`	Correlation coefficients
`cov`	Covariance matrix
`max`	Maximum value
`min`	Minimum value
`mean`	Average
`median`	50th percentile
`mode`	Most frequent value
`std`	Standard deviation
`var`	Variance
Visualization	
`bar, barh, bar3, bar3h`	Bar graphs in 2-D and 3-D
`hist`	Frequency histogram
`pie, pie3`	2-D and 3-D pie charts

TABLE 3.2

List of Functions for Describing Data in the Statistics Toolbox

Statistics	
`corr`	Linear or rank correlation with p-values
`geomean`	Geometric mean
`harmmean`	Harmonic mean
`iqr`	Interquartile range
`kurtosis`	Kurtosis or curvature
`mad`	Mean absolute deviation
`moment`	Central moment—all orders
`percentile`	Percentile—specify percents
`quantile`	Quantile—specify probability
`skewness`	Skewness
`zscore`	Number of standard deviations above or below the mean
Visualization	
`boxplot`	Display boxplots
`histfit`	Histogram with density overlayed
`normplot`	Normal probability plot
`probplot`	Probability plot
`qqplot`	Quantile-Quantile plot

Chapter 4

Probability Distributions

We must become more comfortable with probability and uncertainty.
Nate Silver (1978–)

As we will see later on, data analysis sometimes requires an understanding of probability concepts and the use of probability distributions. The Statistics Toolbox has functionality that will handle many parametric probability density and distribution functions. These functions can estimate distribution parameters using the observed data, generate random variables, and more. In this chapter, we describe the parametric probability distributions available in MATLAB®, discuss how to estimate the parameters of the distributions, and illustrate the functions that generate random variables.

4.1 Distributions in MATLAB®

We discussed some aspects of data distributions in Chapter 3, when we calculated various descriptive statistics and looked at how the data are distributed using histograms, Q–Q plots, and boxplots. Those distributions are a description of the random sample and can be thought of as *empirical*. Another type of distribution is referred to as *theoretical*, and these are derived using mathematical principles or assumptions made about the underlying population. We can use these distributions to obtain a probability that some event will occur. In data analysis, we often start with the observed data (an empirical distribution) and end up using a theoretical distribution to make statistical inferences.

Theoretical probability distributions can be categorized as discrete or continuous, univariate or multivariate, and parametric or nonparametric. In parametric distributions, we find values or *parameters* that govern certain aspects of the distribution, such as location and shape. It is a characteristic of a distribution that distinguishes it from others in the same parametric family. For example, the normal distribution has two parameters—the mean and the variance. There are an infinite number of possible normal distributions, which are distinguished by these two values. In Table 4.1, we list some distributions and their parameters.

In statistical analysis, the parameter(s) of a distribution representing an observed phenomenon are usually taken to be unobservable. However, we use data from our sample to estimate them. These estimates are called *statistics*, and they are defined as functions of our random variables. We described several statistics in the previous chapter, such as the mean, the median, the variance, and the correlation coefficient.

In this chapter, the main focus will be on univariate parametric theoretical distributions, but it will also include some of the multivariate and nonparametric distributions that are available in MATLAB. We start off our discussion by presenting some background information on continuous and discrete probability distributions.

4.1.1 Continuous Distributions

A *continuous variable* is one that can take on any value within an interval of real numbers. As an example, recording the temperature in decimal degrees Fahrenheit would produce a continuous variable. A probability distribution for a continuous random variable will assign a probability to the event that a variable value falls within some specified range of continuous values.

There are two types of probability functions associated with continuous random variables—the cumulative distribution function (CDF) and the probability density function (PDF). The *probability density function* provides the likelihood that the random variable will take on a particular value. The area under the curve defined by the density function must be equal to one, in order for it to be a valid PDF. We obtain the probability that a random variable falls within an interval of real numbers by calculating the area under the PDF over the given range.

A similar procedure produces the *cumulative distribution function*, which is defined as the probability that a real-valued random variable will take on a value less than or equal to a given value a. Thus, this is the area under the PDF from the minimum value of the random variable—typically negative infinity—to a. Notationally, the CDF is given by

$$F_X(a) = \int_{-\infty}^{a} f(x)dx,$$

where X is a random variable, and $f(x)$ is the PDF.

Fortunately, the MATLAB Statistics Toolbox takes care of the mathematical details for us when we use its functions for probability distributions. A partial list of the available continuous univariate distributions is provided in Table 4.1. The first column of the table lists the distribution name, and the

TABLE 4.1

Some Continuous Distributions in the Statistics Toolbox

Distribution Name	Parameters[a]	Root Word	Add to Root Word for Operations
Beta	**a, b**	**beta**	**pdf, cdf, inv, stat, fit, like, rnd**
Chi-Square	**df** (degrees of freedom)	**chi2**	**pdf, cdf, inv, stat, rnd**
Exponential	**mu** (mean)	**exp**	**pdf, cdf, inv, stat, fit, like, rnd**
Extreme Value	**mu** (mean log value) **sigma** (spread log value)	**ev**	**pdf, cdf, inv, stat, fit, like, rnd**
F	**df1, df2** (degrees of freedom)	**f**	**pdf, cdf, inv, stat, rnd**
Gamma	**a** (shape) **b** (scale)	**gam**	**pdf, cdf, inv, stat, fit, like, rnd**
Lognormal	**mu** (mean) **sigma**(spread)	**logn**	**pdf, cdf, inv, stat, fit, like, rnd**
Normal (Gaussian)	**mu** (mean) **sigma** (spread)	**norm**	**pdf, cdf, inv, stat, fit, like, rnd**
Rayleigh	**b**	**rayl**	**pdf, cdf, inv, stat, fit, rnd**
Student's *t*	**v** (degrees of freedom)	**t**	**pdf, cdf, inv, stat, rnd**
Uniform	**a, b** (defines interval)	**unif**	**pdf, cdf, inv, stat, fit, rnd**
Weibull	**a** (scale) **b** (shape)	**wbl**	**pdf, cdf, inv, stat, fit, like, rnd**

[a]The parameters are arguments to the functions that specify the distribution characteristics, and most have a default value. Add the phrase in the fourth column to the root word to run the desired operation. For example, use **unifrnd** to generate uniform random variables in a specified interval. Not all operations require parameter arguments; see **help functionname** for more information on what is needed. For example use **help normcdf** for information on the normal CDF function.

second column lists the parameters that are needed as arguments for some of the functions. The third column shows the root word that MATLAB uses to specify the functions relating to that particular distribution. The remaining column in the table shows the letters that are added to the root function name to indicate what operation will be performed. See Table 4.2 for details.

As an example, the root word for an exponential distribution is **exp**. To evaluate the probability density function, we add **pdf** to the base word, producing **exppdf**. Similarly, we can add **rnd** to the root word, producing **exprnd**, which will generate exponentially distributed random variables.

The *inverse function* (or **inv** in the table) returns the inverse of the CDF for the given distribution. This corresponds to the value of x that produces a given area to the left. This is sometimes called the *quantile function*. For a continuous distribution and a given $q \in [0, 1]$, this is the unique value of x, such that

$$F(x) = q.$$

To be more explicit, we could write the quantile function as

$$F^{-1}(q) = x.$$

Adding the letters **stat** to the distribution name invokes a MATLAB function that returns the mean and variance for the distribution with given parameters. Calling the **fit** operation will estimate the parameters for the distribution using the observed data and the maximum likelihood approach. The **like** option will calculate the negative log-likelihood evaluated at the given parameters and the data. Likelihood functions are used as objective functions for parameter estimation.

We often need to generate random variables for data analysis. For example, we can use computer simulations for hypothesis testing and for estimating confidence intervals. (See Chapter 5 for information on confidence intervals and hypothesis testing.) We can also use random number generation to assess the distribution of our data (to be shown in a later section of this chapter). To generate a set of random variables from a distribution given in Table 4.1, we add the phrase **rnd** to the root word. Thus, we can generate a row vector of n normal random variables with a given mean (**mu**) and standard deviation (**sigma**) by using

```
normrnd(mu,sigma,1,n)
```

4.1.2 Discrete Distributions

As we stated earlier, probability distributions can also be discrete. A *discrete* random variable is one that takes on a value in a countable set of numbers. These variables are usually counts of something in most discrete probability

TABLE 4.2

Codes for Probability Distributions

`cdf`	Cumulative distribution function	`pdf`	Probability density function
`fit`	Estimate parameters	`rnd`	Generate random variables
`like`	Negative log-likelihood	`stat`	Mean and variance
`inv`	Quantile function		

distributions that data analysts encounter. For example, we might have the number of computers owned by a family or the number of typographical errors in a document. If a variable value can correspond to a number in a countable or finite set, then the variable has to be discrete.

We have a ***probability mass function*** instead of a density function when we work with discrete variables. The mass function assigns a probability to the event that a random variable will take on specific discrete values from the domain. Note that the acronym PDF is sometimes used to refer to the probability mass function as well as the probability density function.

The cumulative distribution function for a discrete random variable has the same definition as the continuous case, but the calculation of it is slightly different. The probability that a discrete random variable takes on a value less than or equal to a given value a is obtained by adding the probabilities of the individual events that satisfy the inequality [Ross, 2012], as shown here

$$F(a) = \sum_{\text{all } x \le a} p(x) \quad \text{with} \quad p(x) = P(X = x).$$

A partial list of discrete distributions available in the Statistics Toolbox is given in Table 4.3. The columns of the table are similar to those for continuous distributions (Table 4.1).

4.1.3 Probability Distribution Objects

Recall from Chapter 1 that MATLAB has some data constructs and functions that follow the object-oriented programming paradigm. We encounter our first example of these special data objects with probability distributions.

The Statistics Toolbox has a ***distribution object class*** for the parametric distributions listed in Tables 4.1 and 4.3. This object class is called **ProbDist**, and the object class provides access to more distributions than are listed in

TABLE 4.3

Some Discrete Distributions in the Statistics Toolbox

Distribution Name	Parameters[a]	Root Word	Add to Root for Operations
Binomial	**n, p**	**bino**	**pdf, cdf, inv, stat, fit, rnd**
Geometric	**p**	**geo**	**pdf, cdf, inv, stat, rnd**
Hypergeometric	**m, k, n**	**hyge**	**pdf, cdf, inv, stat, rnd**
Negative binomial	**r, p**	**nbin**	**pdf, cdf, inv, stat, fit, rnd**
Poisson	**lambda**	**poiss**	**pdf, cdf, inv, stat, fit, rnd**
Uniform (discrete)	**N**	**unid**	**pdf, cdf, inv, stat, rnd**

[a]The parameters are arguments to the functions that specify the distribution characteristics, and most have a default value. Add the phrase in the fourth column to the root word to run the desired operation. For example, to obtain parameter estimates and confidence intervals of a binomial distribution based on a data set, use **binofit**. Not all operations require parameter arguments; see the **help** for more information on what is needed. So, to search for information on the hypergeometric probability mass function, use **help hygepdf**.

this book. See the *Statistics Toolbox™ User's Guide* from The MathWorks, Inc. for more information and for a complete list of distributions.

These probability distribution objects provide a simple way to work with data and distributions. Additionally, there are methods or functions that work on the distribution objects that correspond to most of the operations that we saw before. These include evaluating PDFs, CDFs, and the quantile function, as well as generating random variables and estimation. We provide a brief introduction to working with distribution objects here.

There are three main ways to generate a parametric distribution object in MATLAB, as shown here. The objects returned from these functions are an instance of the **ProbDist** class.

1. Use the **fitdist** function to get a probability distribution object **pd** that is fitted to a given distribution (**distname**) using the **data**. The basic syntax is

   ```
   pd = fitdist(data, 'distname')
   ```

2. The function **makedist** will create a probability distribution object for the distribution (with default parameters) given by **distname**. Default parameters are used with this syntax:

   ```
   pd = makedist('distname')
   ```

3. The function **ProbDistUnivParam** produces an instance of a univariate probability distribution specified by **distname** and the parameters **params**. This object is actually a subclass of **ProbDist**.

```
pd = ProbDistUnivParam('distname', params)
```

A partial list of methods that can be applied to these objects is given in Table 4.4. These are functions that perform the specified operation on a probability distribution object. These work as standalone functions. For example, say we have a distribution object called **nrmdist** that we created by using the **fitdist** function. Then, we find a $100 \times (1 - \alpha)\%$ confidence interval for the estimated parameters using this function call

```
ci = paramci(nrmdist,alpha)
```

As another example, if we want the CDF at the values in **x**, then we can use

```
ncdf = cdf(nrmdist,x)
```

TABLE 4.4

Some Methods for Distribution Objects in the Statistics Toolbox

cdf	Return the cumulative distribution function
icdf	Return the inverse CDF
iqr	Interquartile range
mean	Mean
median	Median
paramci	Confidence intervals for parameters
pdf	Return the probability density function
random	Generate random numbers
std, var	Standard deviation and variance

NOTE: Some of the functions are not available for all subclasses of probability distribution objects or distributions. As we saw in Chapter 3, some of the functions (e.g., **mean** and **median**) work on standard MATLAB objects, such as matrices and vectors, in addition to **ProbDist** objects.

4.1.4 Other Distributions

So far, we have been talking about *univariate parametric* distributions. In addition to these, there are other types of probability distributions that are worth mentioning, such as multivariate and nonparametric distributions. MATLAB has several multivariate parametric distributions, but we mention only two of the continuous distributions here. These are the multivariate normal and the multivariate Student's *t*. There is one nonparametric distribution supported by MATLAB, which is the kernel density estimator for univariate data.

The *multivariate normal distribution* is an extension of the univariate case, where each of the *p* dimensions (or variables) is a univariate normal.

Note that the multivariate normal distribution reverts to the univariate version that we saw earlier when $p = 1$.

To characterize a multivariate normal, we have to specify a p-dimensional mean vector and a $p \times p$ covariance matrix. This matrix must be symmetric and positive definite. A matrix is *symmetric* if it is equal to its transpose (rows become columns in the matrix transpose). In other words, the elements are symmetric across the diagonal. A matrix is *positive definite* if its eigenvalues are greater than zero, which means it is invertible. See Chapter 7 for information on eigenvalues.

We saw an example of a multivariate normal distribution earlier, where the individual variates were uncorrelated (see Figure 2.3). In this case, the off-diagonal elements of the covariance matrix are zero. We illustrate the correlated case in the examples given later in this chapter. The functions used for the multivariate normal are shown in Table 4.5.

TABLE 4.5

Additional Distributions in the Statistics Toolbox

mvnpdf	Multivariate normal PDF
mvncdf	Multivariate normal CDF
mvnrnd	Generate multivariate normal random variables
mvtpdf	Multivariate Student's t PDF
mvtcdf	Multivariate Student's t CDF
mvtrnd	Generate multivariate t random variables
ksdensity	Kernel density—arguments include options for PDF, CDF, inverse CDF, and fitting

Like the multivariate normal, the multivariate Student's t distribution is a generalization of the univariate distribution. The multivariate t distribution is parametrized by a degrees-of-freedom parameter and a correlation matrix. It is a distribution for multi-dimensional correlated variables, each of which is a univariate Student's t distribution. The degrees of freedom is a single positive value (or scalar), and the correlation matrix is $p \times p$ positive definite.

Describing the details of kernel densities is beyond the scope of this book. We summarize the steps here, but the interested reader is referred to Scott [1992] for more information. The kernel density estimate of a PDF is obtained by centering a kernel function at each observation and evaluating it over the domain. The function is then scaled by the specified bandwidth, and the final estimated PDF is calculated as the average of the n kernel density functions centered at x. The kernel function could be any valid probability density function, but the normal is typically used.

Example of Kernel Density Estimation

We illustrate the concept of a kernel density estimate in Figure 4.1, and the following MATLAB code shows how we created the plot. We first generate a sample of size $n = 10$ standard normal random variables. Next, we set the bandwidth h for the kernel density using the normal reference rule [Scott, 1992]. We then loop through each of the data points and evaluate a normal PDF using the observation as the mean and the bandwidth h for the standard deviation. Options for the bandwidth are discussed in more detail later in this section.

Each individual kernel function is scaled, and the functions are summed to produce the final estimated density function. We plot the kernel functions and the resulting PDF in Figure 4.1.

```
% Generate a set of normal random variables.
% The mean is 0 and sigma is 1.
n = 10;
x = normrnd(0,1,1,n);

% Get a grid of points to evaluate density.
pts = linspace(-4,4);

% Set up a vector for our estimated PDF.
kpdf = zeros(size(pts));

% We need a bandwidth for the kernels.
% This is the normal reference rule [Scott, 1992].
h = 1.06*n^(-1/5);

% Hold the plot, because we will add
% a curve at each iteration of the loop.
hold on
for i = 1:n
    % Use the normal kernel, noting that
    % the mean is given by an observation
    % and the standard deviation is h.
    f = normpdf(pts, x(i), h);
    % Plot the kernel function for i-th point.
    plot(pts, f/n);
    % Keep adding the individual kernel function.
    kpdf = kpdf + f/n;
end

% Plot the kernel density estimate.
plot(pts, kpdf)
title('Kernel Density Estimate')
xlabel('X')
ylabel('PDF')
hold off
```

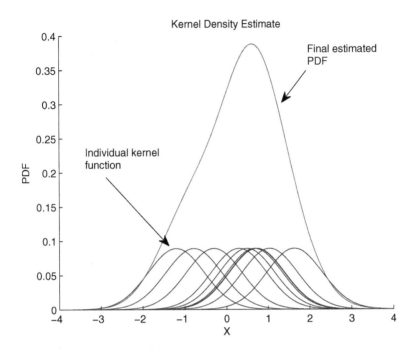

FIGURE 4.1

This is an illustration of the kernel density estimation method. We generated ten standard normal random variables with a mean of zero and a standard deviation of one. Next, we evaluated the kernel function centered at each of the ten points. In this example, the normal kernel was used, where the observation is the mean, and the user-specified bandwidth is the standard deviation. Each kernel function is scaled. The individual curves (or kernel functions) are added to produce the final estimated PDF. Note that the final PDF will have a higher density where there are more data points because there are more kernel functions in those areas.

Kernel densities can be thought of as a type of smoothing operation. We saw an example of a kernel density estimate in Figure 3.6. The smooth curve superimposed over the histogram was obtained using a kernel density. There is a parameter that the user can provide for kernel density estimation. This is the ***bandwidth***, and it controls the amount of smoothing. Larger values of the bandwidth give smoother curves, and smaller values produce curves that fluctuate more. This is analogous to what happens when we change the number of bins in histograms; see Figure 3.3. We will explore this idea in the examples.

Fortunately, we do not have to worry about the details of calculating a kernel density estimate. There is a MATLAB function for kernel densities, which is called **ksdensity**. Using different arguments for this function will return either the probability density function, the cumulative distribution function, or the inverse CDF. This basic call to the function will return the estimated kernel density PDF using the default bandwidth at equally spaced points over the range of **x**. We show it here

$$kpdf = ksdensity(x)$$

To get the inverse CDF based on the kernel density approach and the data in **x**, use

$$kicdf = ksdensity(x,pts,'function','cdf')$$

This returns the estimated CDF evaluated at the points in the vector **pts**. The other operations work in a similar manner—just specify the desired argument and pair it with **'function'**.

4.1.5 Examples of Probability Distributions in MATLAB®

This first example shows how to get the probability density functions for two univariate distributions—the normal and the Student's *t*. We get the density functions for both of these distributions, and we plot them in Figure 4.2. Both of these are symmetric with one mode (or bump). However, the tails of the Student's *t* distribution are usually fatter than the tails of the normal. The *t* distribution shown in Figure 4.2 has one degree of freedom. The Student's *t* will approach the normal, as the degrees of freedom get large. The following MATLAB code generates the PDFs in the plot.

```
% First, generate some points for the domain.
pts = linspace(-4,4);

% Obtain a standard normal PDF.
nrmpdf = normpdf(pts,0,1);

% Obtain a Student's t with v = 1.
stpdf = tpdf(pts, 1);

% Plot the two PDFs.
plot(pts,nrmpdf,'-',pts,stpdf,'-.')
axis([-4 4, 0 0.42])
title('Normal and Student''s t Distributions')
xlabel('x')
ylabel('PDF')
legend('Normal','Student''s t')
```

Our next example shows how to get the PDF and the CDF of a discrete random variable. We will use the Poisson distribution to illustrate the MATLAB functionality. The Poisson distribution is used in applications where the random variable represents the number of times an event occurs in a given interval of some unit. This unit can be time, distance, area, or other entity of interest. Examples of such applications include the number of

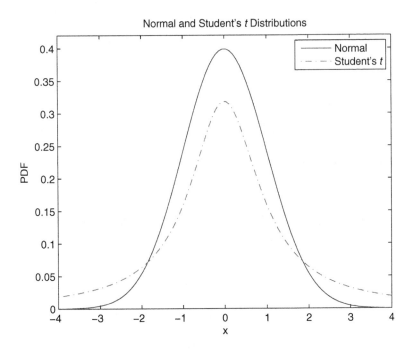

FIGURE 4.2
The solid curve corresponds to a standard normal probability density function, and the dot-
ted curve is the Student's t with one degree of freedom. The Student's t distribution has the
same general shape (symmetric, one mode) as the normal, but it has fatter tails. However,
it will approach a standard normal as the degrees of freedom get larger.

children in a household, the number of books on a shelf, or the number of cars
passing a particular point in an hour.

A Poisson distribution is characterized by one parameter, called λ or
lambda (see Table 4.3). A Poisson random variable can take on values from
$0, 1, 2, \ldots$ The parameter λ specifies both the mean and the variance of the
distribution. In Figure 4.3, we show both the PDF and the CDF for a Poisson
distribution with $\lambda = 3$. The MATLAB code to create them is shown below.

```
% First specify the parameter for the Poisson.
lambda = 3;

% Get the points in the domain for plotting
pts = 0:10;

% Get the values of the PDF
ppdf = poisspdf(pts,lambda);
```

```
% Get the values of the CDF
pcdf = poisscdf(pts,lambda);

% Construct the plots
subplot(2,1,1)
plot(pts, ppdf, '+')
ylabel('Poisson PDF')
title('Poisson Distribution \lambda = 3')
subplot(2,1,2)
stairs(pts, pcdf)
ylabel('Poisson CDF')
xlabel('X')
```

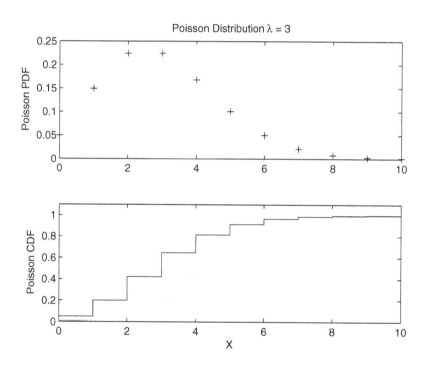

FIGURE 4.3
These are the PDF and the CDF for a Poisson distribution with parameter $\lambda = 3$.

This next example shows how to get a PDF for the multivariate Student's *t* distribution. If we ask for **help mvtpdf** at the command line, then we see that there are two parameter arguments—the correlations and the degrees of freedom. We first set these parameters and the grid for evaluating the PDF.

```
% Example - Multivariate (2-D) Student's t
% First, set the parameters.
corrm = [1, 0.3 ; 0.3, 1];
df = 2;

% Get a grid for evaluation and plotting
x = linspace(-2,2,25);
[x1,x2] = meshgrid(x,x);
X = [x1(:), x2(:)];
```

Now, we are ready to generate the values of the PDF and plot them. The PDF is shown in Figure 4.4.

```
% Evaluate the multivariate Student's t PDF
% at points given by X.
pmvt = mvtpdf(X, corrm, df);

% Plot as a mesh surface.
mesh(x1,x2,reshape(pmvt,size(x1)));
xlabel('X1'), ylabel('X2'), zlabel('PDF')
title('2-D Student''s t Distribution')
```

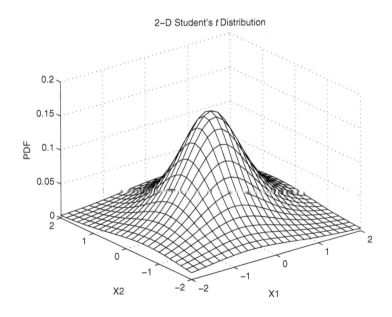

FIGURE 4.4
This shows a mesh surface of a 2-D Student's t distribution with a correlation given by 0.3 and 2 degrees of freedom.

The final example in this section uses the kernel density approach to estimate the PDF of the **earth** data. We will also show the effect of different bandwidths on the resulting estimate. Here are the steps to generate the PDFs shown in Figure 4.5. We use two different bandwidths, both of which are based on the normal reference rule [Scott, 1992]. The first bandwidth **h1** uses the standard deviation, and the second bandwidth **h2** uses the IQR. Recall that the **earth** data had some potential outliers. So, the second bandwidth might be more suitable, because it is based on a robust estimate of scatter.

```
% First load the data into the workspace.
load earth
n = length(earth);

% Set two bandwidths.
% Normal reference rule - standard deviation.
h1 = 1.06*std(earth)*n^(-1/5);

% Normal reference rule - IQR.
h2 = 0.786*iqr(earth)*n^(-1/5);

% Get a domain for the for the PDF.
[kd1,pts] = ksdensity(earth,'bandwidth',h1);
[kd2,pts] = ksdensity(earth,'bandwidth',h2);
plot(pts,kd1,pts,kd2,'-.')
legend('h1 = 0.1832','h2 = 0.1263')
title('Kernel Density Estimates - 2 Bandwidths')
ylabel('PDF'), xlabel('Density of Earth')
```

4.1.6 disttool for Exploring Probability Distributions

MATLAB provides three GUIs to help you manage probability distributions, and we describe the first one here. This is called the Probability Distribution Function Tool. You start the interface by typing **disttool** at the command line. Essentially, this allows you to explore the PDF and CDF of most distributions in the Statistics Toolbox. Here is a description of the interactive capabilities in **disttool**.

- **Distribution Type**: A drop-down menu at the top of the tool includes a list of supported probability distributions. Most of the distributions listed in Tables 4.1 and 4.3 are included.
- **Function Type**: This drop-down menu provides options to display the PDF or the CDF for the given distribution.
- **Function Value**: There is an editable text box on the vertical axis of the CDF. As an example, you can enter a value there, which can be used to find the quantile or critical value.

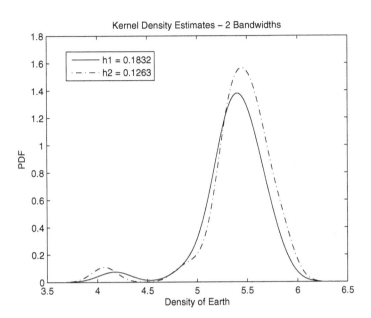

FIGURE 4.5
These are kernel density estimates of the PDF for the **earth** *data.*

- **X Value**: A similar editable text box is on the horizontal axis of the CDF and the PDF. It is also mapped to the red reference lines.
- **Reference Lines**: These are draggable reference lines that are shown as a red dash-dot cross-hair (CDF) or a vertical line (PDF).
- **Parameter Controls**: The bottom of the **disttool** GUI has inputs to control the values of the parameter that correspond to the chosen distribution. These interactive inputs include editable text boxes, as well as sliders. The shapes of the distributions change as one moves the sliders. Note that the upper and lower bounds can be changed in most cases.

Screen shots of the **disttool** interface are shown in Figure 4.6. The upper panel shows the PDF of the F distribution, which is used quite often in hypothesis testing. We see that there are two parameters for this distribution that correspond to two degrees of freedom.

We usually have to find the critical value of some distribution (e.g., Student's t or the F distributions) in many applications of statistical inference. The screenshot in the bottom of Figure 4.6 shows how we can get this value using the **disttool**. Here, we see the CDF of the Student's t distribution for $df = 19$ (degrees of freedom), corresponding to a sample size of $n = 20$. We enter or set these parameters based on our application. Say we want to construct a 95% confidence interval for some estimated parameter, which

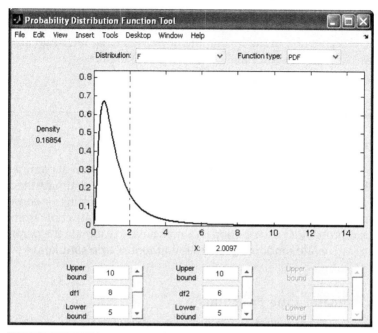

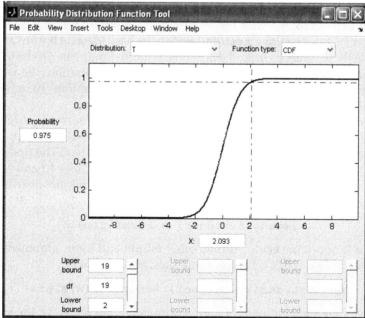

FIGURE 4.6
The upper screenshot shows the **disttool** *interface for the PDF of an F distribution, and the bottom one displays the CDF of the Student's t. There are drop-down menus to change the type of function and distribution.*

requires the value of x that provides an area to the left of 0.975 for the right end of the interval under the Student's t distribution. We can get this by entering 0.975 in the text box next to the vertical axis, and we see the desired quantile or value of x in the box on the horizontal axis. We could also drag the cross-hair until it shows the value we are looking for.

4.2 Parameter Estimation

Many tasks in statistical data analysis require the estimation of parameters of probability distributions. For example, we might need to estimate the means of two data sets and compare them using hypothesis testing or confidence intervals. In another application, we might assume the data come from some theoretical probability distribution. We then have to estimate the parameters in order to generate random variables for a Monte Carlo simulation.

4.2.1 Command Line Functions

We have already described some MATLAB functions to estimate parameters in Section 4.1, and we expand on them here. We also touched on parameter estimation in Chapter 3 with the discussion on descriptive statistics. As you might recall, the functions **mean** and **std** in the base MATLAB software will calculate these statistics from the observed data, and they can also be used as functions to estimate the parameters of a normal distribution. We will see that the functions in the Statistics Toolbox offer more options than what is available in base MATLAB.

Parameter estimation for probability distributions is sometimes referred to as *distribution fitting*. There are three main command-line functions or approaches for fitting distributions in the Statistics Toolbox. The first one is to use the distribution specific functions, as outlined in Tables 4.1 and 4.3. For example, to get the mean and standard deviation for the normal distribution, use this function call:

$$\texttt{[meanest,stdest] = normfit(x)}$$

where **x** is our data vector. Adding other output and input arguments will return the $100(1 - \alpha)\%$ confidence intervals, as shown here:

$$\texttt{[meanest,stdest,mci,sci] = normfit(x,alpha)}$$

If we keep the same number of output arguments, but leave out the **alpha**, then MATLAB will return the 95% confidence interval, as the default.

Estimating parameters and confidence intervals for other distributions work in a similar manner, as discussed in Section 4.1. For instance, to estimate the parameter for the Poisson distribution, we would use the function **poissfit**. It is always good practice in any analysis to look for **help** on a particular function and see what options are available.

The next function one could use is called **mle**. This function uses the maximum likelihood approach for estimating a parameter. The *maximum likelihood method* seeks the value of the distribution parameter that maximizes the likelihood function, given the observed data. The **mle** function supports most of the distributions available in the Statistics Toolbox. Use **help mle** at the command line to see a list of them. An advantage of the **mle** function is that it allows you to supply the PDF for any parametric distribution you can write, even if it is not directly supported by MATLAB.

The basic syntax for **mle** is

```
est = mle(X,'distribution','name'),
```

where **name** specifies the distribution, and **X** is the data. See the **help** documentation for options. A $100 \times (1 - \alpha)$ % confidence interval is obtained using

```
[est,estci] = mle(x,'distribution','name','alpha',alpha)
```

This syntax will return vectors with the requested estimates. The output variable **est** has the estimates, and the confidence intervals are in **estci**.

The functions we just described return MATLAB vector data objects. The **fitdist** function will estimate parameters for a given distribution, and it returns a **ProbDist** object that represents the fitted distribution. The syntax is simple:

```
PDest = fitdist(x,'distname')
```

See **help** on **fitdist** for a list of distribution names. You also have the option of using **'kernel'** to get a univariate kernel density estimate. Once you have the **ProbDist** object, you can use any of the methods for this type of object (see Table 4.4).

As an example, to get the confidence interval for the estimated parameters, use the following function call:

```
estci = paramci(PDest)
```

4.2.2 Examples of Parameter Estimation

Our first examples illustrating parameter estimation will use the **earth** data. We know from the plots in Chapter 3 that the **earth** data do not appear to be normally distributed. However, we can still estimate the parameters for a normal distribution from the data, using the following MATLAB code.

```
% Load the earth data
load earth

% Estimate the parameters of a normal distribution.
```

```
% Also ask for a 90% confidence interval
[mu1,sig1,muci1,sigci1] = normfit(earth,0.9);
```

We asked for a 90% confidence interval. The estimated mean is 5.4197, and the confidence interval is [5.4117 , 5.4296]. The estimated standard deviation is 0.3389, with a 90% confidence interval of [0.3372 , 0.3489]. We display the results from **normfit** here using the **display** function.

```
display(mu1)

% The following is the estimated mean.
mu1 =

    5.4197

display(muci1)

% Here is the estimated confidence interval.
muci1 =

    5.4117
    5.4276

display(sig1)

% The following estimated standard deviation is shown.
sig1 =

    0.3389

display(sigci1)

% This is shown in the window.
sigci1 =

    0.3372
    0.3489
```

The estimated parameters from **normfit** give the same results as **mean** and **std**, as shown below.

```
% Get the mean from the data.
mean(earth)

% This is returned.
ans =
    5.4197
```

```
% Get the standard deviation
std(earth)
% This is returned in the window.
ans =
    0.3389
```

Next, we use **fitdist** to create a **ProbDist** object by estimating the parameters of the normal distribution and using the **earth** data. We left off the semi-colon, so the characteristics of the **ProbDist** object are displayed in the command window. We see the estimates of the parameters and a 95% confidence interval. Note that the interval given below is wider because we specified a larger confidence—95% instead of 90%. However, the estimated mean and standard deviation are the same.

```
% Fit a normal distribution to the data.
pdfit = fitdist(earth, 'normal')

% This is shown in the window.
pdfit =

   NormalDistribution

   Normal distribution
        mu =  5.41966      [5.29075, 5.54856]
     sigma = 0.338879      [0.268928, 0.458318]
```

We can extract the confidence intervals using the **paramci** function. This is one of the methods we can use with a **ProbDist** object.

```
% Display the confidence intervals only.
% The intervals are given in the columns.
paramci(pdfit)

ans =

    5.2908     0.2689
    5.5486     0.4583
```

Now, let us try a nonparametric distribution for the **earth** data and fit a kernel density using **fitdist**.

```
% Fit a kernel density to the earth data.
ksfit = fitdist(earth, 'kernel');
```

These are the results. Note that parameters and confidence intervals are not returned in this case.

```
ksfit =

   KernelDistribution
```

```
Kernel = normal
Bandwidth = 0.128129
Support = unbounded
```

We can get the PDF over the range of the data for both of these fitted distributions, as follows:

```
% Get a set of values for the domain.
pts = linspace(min(earth)-1,max(earth)+1);

% Get the PDF of the different fits.
pdfn = pdf(pdfit,pts);
pdfk = pdf(ksfit,pts);
```

It would be interesting to plot both of these estimated PDFs and to compare them. These are shown in Figure 4.7, where we also plotted the observations as points along the horizontal axis. This shows where there are more data points whose kernel functions pile up to produce the bumps. As expected, the estimated normal PDF gives just one mode (or bump) and misses the small bump around 4. The nonparametric PDF based on the kernel density estimation does not miss the second bump.

```
% Plot both curves
plot(pts,pdfn,pts,pdfk,'--')

% Plot the points on the horizontal axis.
hold on
plot(earth,zeros(1,n),'+')
hold off
legend('Normal Fit','Kernel Fit')
title('Estimated PDFs for the Earth Data')
xlabel('Density of the Earth')
ylabel('PDF')
```

Our final example in this section shows how to fit a multivariate normal distribution using the Fisher's iris data. We estimate a bivariate PDF for two of the variables of the *Virginica* species to make it easier to visualize. The Statistics Toolbox does not have a specific function for fitting or estimating multivariate distributions, as we have in the univariate case. However, we can estimate the parameters of the multivariate normal using the functions **mean** and **cov**, both of which are in base MATLAB.

We are going to use the first and third variables of the *Virginica* species for this example, because they seem to have a correlated relationship based on the scatter plot matrix shown in the top panel of Figure 4.8. Recall from Chapter 2 that a scatter plot matrix is a display of all pairwise scatter plots of a data set, where the plots are laid out as a table or matrix of plots. We can see

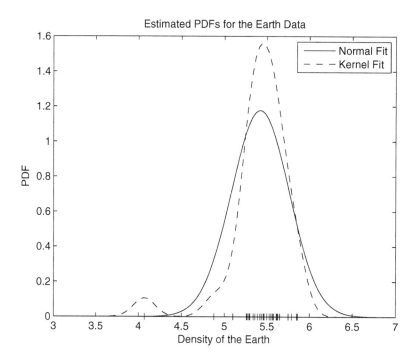

FIGURE 4.7

*Here we have two estimated distributions for the **earth** data. The solid curve represents the fit based on the normal distribution, and the dashed line corresponds to a fit using the kernel density approach. We also include a 1-D scatter plot of the data along the horizontal axis. This gives us an indication of where the data pile up to produce bumps or modes. We know that the normal distribution has only one bump, and thus, will miss the small bump around $x = 4$. The nonparametric kernel density approach does not miss the second bump.*

the scatter plot of the sepal length (vertical axis) versus petal length (horizontal axis) in the scatter plot located in the first row and the third column of Figure 4.8. Similarly, the plot of petal length (vertical axis) against sepal length (horizontal axis) is shown in the third row and first column.

```
% Example of multivariate normal using the iris data.
load iris

% Construct a scatter plot matrix.
% This function is in base MATLAB.
plotmatrix(virginica)
title('Iris Virginica')
```

Now, let's fit the distribution. Here is the code we use to do it.

```
% Extract the variables from the Virginica species.
X = virginica(:,[1,3]);
```

```
% Estimate the mean.
mu = mean(X);

% Estimate the covariance.
cv = cov(X);
```

The next step is to create the grid of points where we will generate values of the estimated PDF.

```
% Establish a grid for the PDF
x1 = linspace(min(X(:,1))-1,max(X(:,1))+1,25);
x2 = linspace(min(X(:,2))-1,max(X(:,2))+1,25);
[X1, X2] = meshgrid(x1,x2);
```

We are now ready to generate the values of the estimated PDF, as shown here.

```
% Use the parameter estimates and generate the PDF.
pts = [X1(:),X2(:)];
Xpdf = mvnpdf(pts,mu,cv);
```

We create a mesh plot to view the results. We show this in Figure 4.8, along with a scatter plot for comparison. This and other three-dimensional plots can be rotated using the rotation button on the Figure window toolbar.

```
% Construct a mesh plot.
mesh(X1,X2,reshape(Xpdf,25,25))
title('Estimated PDF for Iris Virginica')
xlabel('Sepal Length'), ylabel('Petal Length')
zlabel('PDF')
```

> **TIP**
>
> **ProbDist** objects have useful property values. Search the documentation for *ProbDistParametric class* for a list of properties. A useful one is **Params**. This returns the estimated parameters for the object.

4.2.3 dfittool for Interactive Fitting

The Statistics Toolbox has a GUI for fitting distributions to data. It is called the Distribution Fitting Tool. Type **dfittool** at the command line to start it, and see Figure 4.9 for an example of the opening display. It has buttons and user interfaces that allow you to import data into the GUI from the workspace, change the display type, create and manage fits, and evaluate the distribution at given points. We briefly describe some of the main tasks.

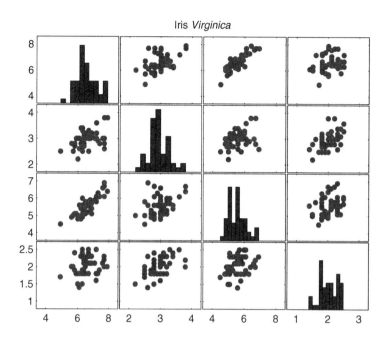

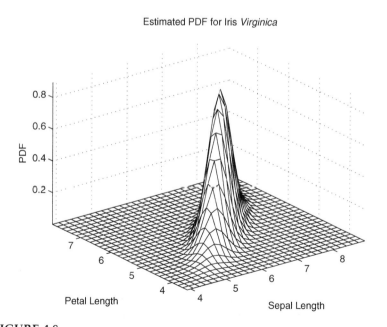

FIGURE 4.8
The top panel is a scatter plot matrix for iris Virginica. *A plot of the first and third variables is in the first row, third column. We estimated the parameters for the multivariate normal and use them to produce the fitted distribution shown in the bottom panel.*

- **Display Type**: You can choose to display the fitted distribution as a PDF, CDF, or inverse CDF (the quantile function). You can also display a probability plot, a survivor function, or a cumulative hazard plot.

- **Data**: This button will open a dialog box, where you can select the data set to use, specify rows or columns for the analysis, preview the data, and create the data set for the GUI by clicking the CREATE DATA SET button.

- **New Fit**: Selecting this button in the main window opens a dialog box, where you can name the fit, select the data from a list of created data sets, and specify the distribution to fit. Push the APPLY button to estimate the distribution.

- **Manage Fits**: This button opens a box where you can display the plot in the main window of the tool, display pointwise confidence bounds in the CDF or inverse CDF plot, edit your fit, delete it, and save it to the workspace.

- **Evaluate Fits**: Pushing this button opens a dialog box, where you can choose points at which to evaluate the fit. You can also calculate confidence intervals for most distribution fits and export the results to the workspace.

- **Exclude**: Select this option if you need to exclude data for some reason. This can be based on setting limits, establishing rules, or selecting them graphically.

The Distribution Fitting Tool works with vector objects. So, only matrices or vectors that are currently in the workspace will appear as options in the Data dialog box menu. The lower panel of Figure 4.9 shows the Data dialog box, after we created a data set using the **earth** data.

Note that there is a pane at the bottom of the **DATA** dialog box that allows you to manage the data sets that you create for this tool. A list of these created data sets will appear, and you have options to view them in a new window, to add confidence bounds for the CDF, to set the bin rules used in the histogram plots, or to delete the fitted distribution. Clicking the SET BIN RULES button in the Data dialog box opens up another GUI that provides several choices for specifying the bin rules. These include the Freedman-Diaconis rule, Scott's rule, setting the number of bins or the width, and more [Scott, 1992].

Once the data sets are imported, you use the NEW FIT button to choose the distribution for fitting and to estimate the parameters. This will create the fit, display it in the main window, and provide a summary of the estimates and other results. You can also save the fit as a **ProbDist** object to the workspace using the SAVE TO WORKSPACE button. In Figure 4.10, we show the main window of the tool, where we have a density histogram, an estimated normal distribution, and a kernel density fit for the **earth** data.

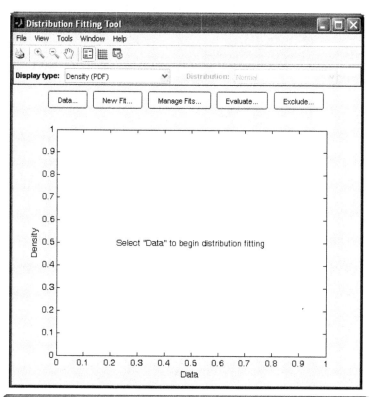

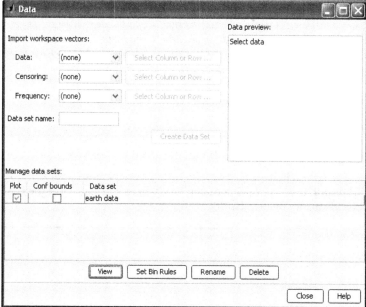

FIGURE 4.9
*These are screen shots of a newly opened **dfittool** session and the Data dialog box.*

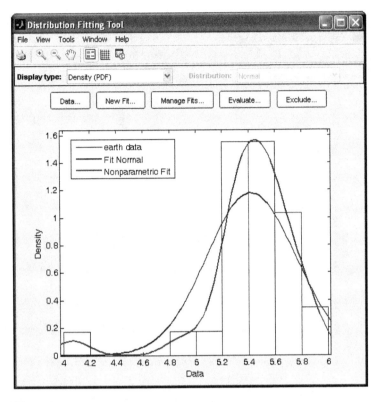

FIGURE 4.10
We estimated the parameters for a normal distribution for the **earth** *data using the Distribution Fitting Tool, and the fit is displayed here. Then, we fit a nonparametric kernel density. Note that the tool also shows the density histogram of the data, which is helpful in assessing the fit. The rules for the histogram bins can be changed using the* **Data** *button.*

<u>TIP</u>

There is an option in the **dfittool** for fitting a nonparametric or kernel density. Choosing this will open a dialog box, where you can select the type of kernel and specify the bandwidth.

4.3 Generating Random Numbers

The need to generate random numbers from probability distributions arises in many statistical methods and data analytic tasks. For example, one could use Monte Carlo computer simulations for statistical inference and the bootstrap method to estimate variances and confidence intervals. In this section, we describe the functions for generating random variables in base MATLAB and the Statistics Toolbox. We conclude with a discussion of a GUI tool for random number generation and some examples.

4.3.1 Generating Random Variables in Base MATLAB®

The base MATLAB package has two functions for generating variables. The first of these functions is called **rand**, and it will generate uniform random variables in the interval $(0, 1)$. The second function is called **randn**, and it generates random variables from a standard normal distribution.

Most algorithms for generating random variables from a given probability distribution start with random numbers from a uniform $(0, 1)$ distribution. It is fairly easy now to get such random numbers using computers, and the ability to generate them is in most computational software packages. It is important to note that these are really pseudorandom numbers, because most algorithms rely on a deterministic approach to generate them. However, the numbers obtained using these algorithms are random enough to suit most of our data analytic tasks. This idea will be illustrated in the examples.

The following function call will generate an $n \times p$ array of uniform $(0, 1)$ random variables:

$$r = \text{rand(n,p)}$$

You can easily use **rand** to generate random variables over an interval (a, b) using this transformation:

$$X = (b - a) \times U + a,$$

where U is a uniform $(0, 1)$ random variable.

A normal probability distribution is used quite often in data analysis and statistics. So, the ability to generate random numbers from this distribution is helpful. We can use the function **randn** in base MATLAB to get random variables from a standard normal distribution. Recall that this corresponds to a normal distribution with a mean of zero and a variance of one. The syntax for this function is similar to **rand**, as shown here

$$r = \text{randn(n,p)}$$

We can generate univariate normal random variables for a given mean μ and variance σ^2 using this transformation

$$X = Z \times \sigma + \mu,$$

where Z is a standard normal random variable.

TIP

There is another function for generating random variables in base MATLAB. It is called **randi**, and it will generate integer random numbers from a discrete uniform distribution. To generate a row vector of 10 integer numbers from 1 to 15, use **randi(15,1,10)**.

4.3.2 Generating Random Variables with the Statistics Toolbox

We already discussed random number generation functions in the Statistics Toolbox. Recall that Tables 4.1 and 4.3 list many probability distributions for continuous and discrete variables. We can generate random numbers from these distributions by attaching the letters **rnd** to the root word specifying the distribution. For example, we can generate variables from the normal, exponential, and Poisson distributions using **normrnd**, **exprnd**, and **poissrnd**, respectively. Look at the **help** for the specific functions for more details, but they mostly follow this structure for the input arguments:

$$r = rootrnd(para1,para2,...,n,p)$$

where *para1, para2*, etc., are the parameters required for the particular distribution, **n** is the number of rows, and **p** is the number of columns.

We also talked about the object-oriented approach for working with probability distributions, where we described the **ProbDist** object. If we have a **ProbDist** object, then we can use the function **random** to get random variables based on the distribution type and parameters given by the object. There is no need to specify the parameters again, when using this type of object. The function call for a **ProbDist** object **pd** is

$$random(pd,n,p)$$

The **random** function can also be used without a **ProbDist** object, as shown here

$$random(distname,param1,param2,...,n,p)$$

where *distname* is a string specifying one of the supported distributions. See **help random** for a list of these distributions.

4.3.3 Examples of Random Number Generation

Our first example shows that the random numbers in MATLAB are obtained according to a deterministic algorithm. Most—if not all—of these methods have a starting point or a seed. We can use the base MATLAB function **rng** to set the seed to a specific value, to the current time, or to the default value. In MATLAB 2014a, the default is to use the Mersenne Twister algorithm with a seed equal to zero. The **rng** function can also be used to specify a different random number generator. You can consult the documentation on **rng** for more information on what random number generators are available.

We are first going to set the seed for the random number generator to some value—say 10. Next, we call **rand** twice, where we generate three random numbers each time.

```
% Set the seed to 10.
rng(10)

% Generate 3 random variables.
r1 = rand(1,3)

% This is shown in the command window.
r1 =
    0.7713    0.0208    0.6336

% Generate 3 more random variables.
r2 = rand(1,3)

% This is what we get in the window.
r2 =
    0.7488    0.4985    0.2248
```

We reset the seed back to 10 and then generate six variables.

```
% Now, set the seed back to 10.
rng(10)

% Generate 6 random variables.
r3 = rand(1,6)

% We get the same 6 values from above.
r3 =
  Columns 1 through 3

    0.7713    0.0208    0.6336

  Columns 4 through 6

    0.7488    0.4985    0.2248
```

The same six random variables have been generated because we reset the seed to the same value of 10. Thus, we see that the random numbers are generated in a deterministic manner. This has an advantage in simulations, where we generate a set of random numbers in each trial. We could set the seed to some given value—perhaps a function of the looping variable—that allows us to regenerate the same data, as long as we know the iteration number.

The next example shows how we can use random number generation to get a Q–Q plot. We used the **earth** data previously, where we estimated the parameters of a normal distribution and plotted it (see Figure 4.10). We also used these data previously to demonstrate Q–Q plots using the **qqplot** function (see Chapter 3). We now show how we can create this type of plot by generating random numbers from the desired theoretical distribution, which is a quick and easy way to get those quantiles. For example, this is the MATLAB code we use to create a Q–Q plot comparing the **earth** data to an exponential distribution.

```
% Create a Q-Q plot of the earth data, where
% we compare it to an exponential distribution.
load earth

% Generate exponential variables.
% Use the mean of the data for the parameter.
rexp = exprnd(mean(earth),size(earth));

% Create the Q-Q plot.
plot(sort(earth),sort(rexp),'o')
xlabel('Data'),ylabel('Exponential')
title('Q-Q Plot of Earth Data and Exponential')

% Add a line that is estimated using quartiles.
% Get the first and third quartiles of the earth.
qeth = quantile(earth,[.25,.75]);

% Now, get the same for the exponential variables.
qexp = quantile(rexp,[.25,.75]);

% Fit a straight line. See Chapter 6 for more details.
p = polyfit(qeth,qexp,1);
pp = polyval(p,[5,max(earth)]);
hold on
plot([min(earth),max(earth)],pp)
```

The plot is shown in Figure 4.11, and we see that the data do not appear to be distributed as an exponential distribution. To help our assessment, we obtain a straight line fit based on the first and third quartiles and using the **polyfit** function. The **polyfit** and **polyval** functions will be covered in Chapter 6.

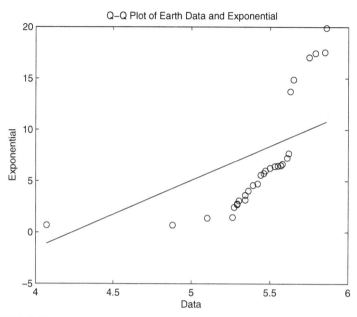

FIGURE 4.11
*We created this Q–Q plot by generating a set of exponential random variables, sorting both sets of data, and constructing a scatter plot. The points do not fall on a line, indicating that the **earth** data are not exponentially distributed.*

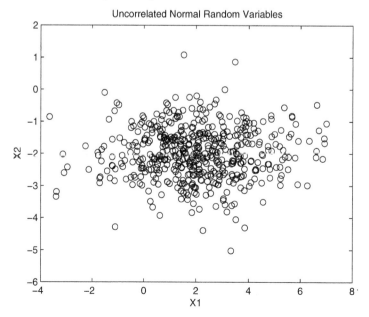

FIGURE 4.12
This is a scatter plot of bivariate normal data. Note that they appear to be uncorrelated.

Our last example shows that we can use the **randn** function to generate multivariate normal variables, where each variable can have a given mean and variance. However, they will be uncorrelated.

The following code generates 500 bivariate data points and produces the scatter plot in Figure 4.12.

```
% Generate two vectors of normal random variables
% with different means and variances.
% First is a variance of 4 and mean of 2.
x1 = randn(500,1)*sqrt(4) + 2;

% Next is a variance of 0.7 and a mean of -2.
x2 = randn(500,1)*sqrt(0.7) - 2;

% Construct a scatter plot.
plot(x1, x2, 'o')
title('Uncorrelated Normal Random Variables')
xlabel('X1'), ylabel('X2')
```

We see that they are approximately uncorrelated from the sample correlation coefficient matrix.

```
corrcoef([x1,x2])

% This is what we have for these random numbers.
ans =

    1.0000    0.0037
    0.0037    1.0000
```

Next, we use the Statistics Toolbox function **mvnrnd** to generate bivariate normal random variables that are correlated. We first have to set up the mean and covariance matrix. These are used as input to the function. The output from the **mvnrnd** is a matrix of multivariate random variables, where the number of columns is determined by the number of elements in the vector of means. The plot is shown in Figure 4.13, and we can easily see that the variables are correlated by the shape of the data cloud.

```
% Generate bivariate correlated random variables.
mu = [2 -2];
covm = [1 1.25; 1.25 3];
X = mvnrnd(mu,covm,200);

% Show in a scatter plot.
scatter(X(:,1),X(:,2))
xlabel('X_1'),ylabel('X_2')
title('Correlated Bivariate Random Variables')
```

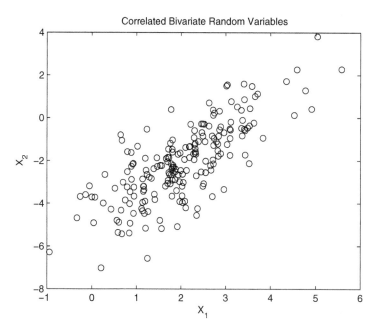

FIGURE 4.13

Here is a scatter plot of the bivariate normal random variables we generated using the function **mvnrnd***. The variables appear to be correlated, as we requested with the input covariance matrix. We can generate multivariate random variables with a specified mean and covariance structure using this function.*

4.3.4 randtool for Generating Random Variables

The Statistics Toolbox has a GUI that will generate random numbers from most of the supported distributions. To start the interface, type in the word **randtool** at the command line. This tool works similarly to the **disttool**, in that it provides an interactive interface that includes text boxes to enter parameters and the number of sample points to generate, sliders to change the parameter values, a menu to select the distribution, and buttons to export the data to the workspace or to resample. There is also a histogram of the data to show the distribution shape.

MATLAB will automatically generate a new sample from the requested distribution, as you change the parameters. So, you can move the slider for the parameters and see what happens in real time. If you want a different sample using the same parameters, then push the **RESAMPLE** button. A screen capture of the **randtool** is given in Figure 4.14. The example shows a histogram of 100 binomial random variables.

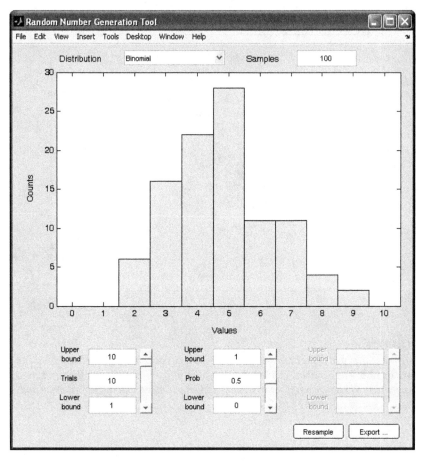

FIGURE 4.14
Here is an example of the **randtool**, *where we generated 100 numbers from a binomial.*

4.4 Summary and Further Reading

In this chapter, we provided an overview of MATLAB functions for handling probability distributions. Most of these are in the Statistics Toolbox, but some functionality for generating uniform (**rand**) and standard normal (**randn**) random variables are in the base MATLAB. We also described three GUIs for distribution fitting, exploring distributions, and random number generation. These are summarized in Table 4.6.

There are many good introductory books on probability distributions and theory. In particular, the book *A First Course in Probability* [Ross, 2012] is at the advanced undergraduate level and is appropriate for professionals and

TABLE 4.6

Probability Distribution GUIs in the Statistics Toolbox

dfittool	Estimate parameters for distributions, visualize fits, evaluate PDFs and CDFs, and more
distool	Explore univariate distributions as PDFs and CDFs
randtool	Generate univariate random variables

students in any discipline. The book *Intuitive Probability and Random Processes using MATLAB* [Kay, 2005] provides details on probability theory and uses MATLAB to illustrate the concepts.

The *Computational Statistics Handbook with MATLAB* [Martinez and Martinez, 2007] is a good resource that has more information on probability distributions and how they arise in many data analytic tasks, such as Monte Carlo experiments, random number generation, model-building, and hypothesis testing.

For information on density estimation using histograms and the kernel density approach—both univariate and multivariate—see David Scott's book *Multivariate Density Estimation: Theory, Practice, and Visualization* [Scott, 1992]. This contains details on the derivation and use of various binwidth rules for nonparametric density estimation. Another excellent resource is *Density Estimation for Statistics and Data Analysis* by Silverman [1986].

There are many books on probability distributions and parameter estimation methods, such as maximum likelihood. We recommend the book *Modern Mathematical Statistics with Applications* [Devore and Berk, 2012]. The authors include examples of data analysis along with the underlying theory to help the reader understand the concepts.

We conclude this chapter by pointing the reader to relevant sections in the MATLAB documentation. You can access the online documentation by going to the RESOURCES section of the desktop ribbon and pushing the HELP button. Click on the MATLAB link to view the sections on the base software. Random number generation is discussed in the *Mathematics ... Statistics and Random Number Generation* section.

Clicking on the Statistics Toolbox link will bring up the documentation on that toolbox. There is a chapter on *Probability Distributions*. This has details on the probability distributions discussed in this chapter, as well as random number generation.

Chapter 5

Hypothesis Testing

Facts are stubborn things, but statistics are pliable.
  Mark Twain (1835–1910)

The first part of this book presented various methods belonging primarily to the realm of descriptive statistics. These included methods for summarizing our data using numbers and visualization techniques. Another part of statistics is concerned with drawing conclusions about the population, based on observed data. This area is called *inferential statistics*, and it includes hypothesis testing and the estimation of confidence intervals. As we will see, it relies on some descriptive statistics we introduced earlier.

  We first briefly introduce the basic concepts of hypothesis testing and confidence intervals. This is followed by commonly used hypothesis tests, such as the z-test and the *t*-test. Next, we describe the bootstrap resampling method for estimating confidence intervals. We conclude this chapter with an introduction to the analysis of variance (or ANOVA), which tests the hypothesis that groups are different in the characteristic of interest.

5.1 Basic Concepts

In this section, we briefly introduce two of the main methods in inferential statistics: hypothesis testing and confidence intervals. With confidence intervals, we estimate the interval we expect contains the true population value of a parameter. In hypothesis testing we make a decision between rejecting or not rejecting some statement about the population [Bhattacharyya and Johnson, 1977; Martinez and Martinez, 2007] based on observed data.

5.1.1 Hypothesis Testing

In hypothesis testing, we start with a statement, which is a conjecture about our population of interest. For example:

- We think a new drug to relieve pain is better than those currently on the market;
- We suspect that the rental cost for a one-bedroom apartment in Washington, D.C., is less than that in Manhattan, New York; or
- A health official believes that high blood pressure is related to age, gender, and weight.

We generally formulate our statistical hypotheses in two parts—the null hypothesis and the alternative hypothesis. The ***null hypothesis*** is the one we would like to test and is represented by H_0. It is called *null* because it typically represents the current status or no effect, and we are investigating departures from it. We also have an ***alternative hypothesis***, and this is denoted by H_1. Rejecting the null hypothesis H_0 leads to the decision to accept or not reject the alternative hypothesis H_1.

In essence, we try to determine if there is significant evidence to reject the null hypothesis in favor of the alternative. When we perform a statistical hypothesis test, we can never know with certainty which hypothesis is true. In this primer, we will use the phrases *accept the null hypothesis* and *reject the null hypothesis* for our decisions resulting from statistical hypothesis testing. We note that the standard language is to *fail to reject the null hypothesis*.

We use the second example above to illustrate these concepts. Denoting the average rent for a one-bedroom apartment in Manhattan as μ_0, we can write the hypotheses as

$$H_0: \qquad \mu = \mu_0$$
$$H_1: \qquad \mu < \mu_0$$

The average rent in Washington, D.C., is represented by μ. The alternative hypothesis is that rent for one-bedroom apartments in Washington, D. C., (μ) is less than the rent in Manhattan (μ_0).

We need to obtain a random sample of data from the population in order to test our hypothesis. Once we have the data, we calculate a ***test statistic*** that will provide some information or insights regarding our hypothesis. Continuing with our example of costs for apartments, we would collect a random sample of rents in Washington, D.C., and use the mean or average rent as our test statistic. If the value of the test statistic is consistent with our null hypothesis, then we accept H_0 (or do not reject H_0). Otherwise, we reject H_0 in favor of the alternative hypothesis.

We need a way to determine when the observed value of a test statistic is consistent with the null hypothesis. To make our decision, we have to know the distribution of the test statistic when the null hypothesis is true. Typically, one derives the distribution for the statistic using theoretical principles.

Let's return to our example. Say we have a random sample comprised of monthly rents for 50 one-bedroom apartments in Washington, D.C. We calculate the mean, and we get a value of $1,750. The average monthly rent for one-bedroom apartments in Manhattan is known to be $1,800. This is higher than the average rent in our sample. However, is the difference *statistically significant*? The mean is a random variable, and it has some variation associated with it. If the variance of the mean (our test statistic) under the null hypothesis is large, then the observed value of $\hat{\mu} = \bar{x} = 1750$ might still be consistent with the null hypothesis H_0.

Our decision in statistical hypothesis testing is based on whether or not our test statistic falls within some region of its probability distribution under the null hypothesis. This is called the ***critical region***. The boundary of the critical region is given by the ***critical value***. This value is the quantile that divides the domain of the distribution into a region where the null hypothesis will be accepted or not accepted.

The critical regions are in the tails of the distribution of our test statistic under the null hypothesis. The location of the critical region depends on the type of alternative hypothesis we have. There are three types of alternative hypotheses that produce three types of critical regions, as listed here.

- **Right-tail test region**: This is the region we have when we are looking for evidence that our test statistic is greater than the null value. The critical region is in the upper tail of the distribution of the test statistic.

- **Left-tail test region**: We have this situation when our alternative hypothesis has a statement that our test statistic is less than the null value. The critical region is in the lower tail of the distribution.

- **Two-tail test region**: In this case, we have two critical values—one for each tail of the distribution. We are interested in evidence that the statistic is different from the null value; it could be significantly greater or lower.

The size of the critical region is determined by the error we are willing to accept in our decision-making. There are two types of errors in hypothesis testing—Type I error and Type II error. The *Type I error* happens when we reject the null hypothesis when it is really true. This is the error that we specify in advance, and it is the error that determines our critical value. The *Type II error* occurs when we do not reject the null hypothesis when it is false.

The *significance level* α of the test is the probability of making a Type I error. The analyst sets the significance level in advance. A typical value is $\alpha = 0.05$. The critical values for the three types of alternative hypotheses are calculated using quantiles, as listed here.

- **Right-tail test value**: The value of x, such that $F(x) = 1 - \alpha$.
- **Left-tail test value**: The value of x, such that $F(x) = \alpha$.
- **Two-tail test value**: The values of x, such that $F(x) = \alpha/2$ (lower critical value) and $F(x) = 1 - \alpha/2$ (upper critical value).

Recall from the previous chapter that a quantile is found using the inverse cumulative distribution function (CDF), which is denoted by $F(x)$. In this case, x denotes the value of a test statistic, and F represents the CDF of the test statistic under the null hypothesis.

Our final concept in hypothesis testing is the ***p-value***. This is the probability of observing a value of the test statistic as extreme as or more extreme (in terms of the alternative hypothesis) than the one that is observed, when the null hypothesis is true. If the p-value $< \alpha$, then we reject H_0.

5.1.2 Confidence Intervals

We first mentioned confidence intervals in Chapter 4, when we discussed how to estimate parameters of a distribution. We now provide some more information on how to construct these intervals.

There are two main ways that we can estimate parameters—as points or intervals. For example, the average value calculated for a data set is a point estimate of the mean. It is just a single number (for univariate data). An alternative approach is to produce an interval estimate, which is a range of values that we assume contains the true population value of a parameter.

This second approach is called a *confidence interval*. We calculate this type of interval using our observed data, as we do with point estimates. Because of this the confidence interval can change depending on the sample. We then assign a level of confidence that reflects the proportion of these intervals that would contain the true population parameter.

The confidence intervals are calculated using a point estimate of a statistic, as shown here.

$$\hat{\theta} \pm E$$

Our parameter of interest is denoted by θ, and we put an upper caret (or hat) above it to indicate that it is an estimate. So, we start with the point estimate of the parameter. We add a quantity E to get the upper bound of the interval, and we subtract E to get a lower bound. E is called the ***margin of error***. It is usually a function of the standard error of the point estimate and a critical value (or quantile).

The confidence level is given by $1 - \alpha$. This means that if we collect '*all possible*' random samples and calculate the confidence interval for each sample, then we would expect $100 \times (1 - \alpha)\%$ of the intervals to contain the true value of the population parameter. This is the same α we had with hypothesis testing, and one of the examples given later will illustrate how the two-tail test is related to a confidence interval.

We will not go into any details of how to calculate these intervals because MATLAB will do that for us, as we saw in Chapter 4. We will see some alternative functions to obtain intervals later in this chapter.

5.2 Common Hypothesis Tests

As we just discussed, statistical hypothesis testing can be used to determine if a particular claim is significant, i.e., if statistical evidence exists in favor of or against a given hypothesis. There are several established hypothesis tests that are available in the MATLAB® Statistics Toolbox, and we describe two of the commonly used ones in this section—the z-test and the t-test.

5.2.1 The z-Test and t-Test

The statistic used in the z-test or the t-test is based on the average $\bar{x}$ of the sample values. However, the tests are different in the assumptions made about the distribution of the random variable that is observed. These assumptions must be considered when choosing a test and interpreting the results. The z-test and the t-test both assume that the data are independently sampled from a normal distribution with mean μ. The z-test assumes that the population standard deviation σ is known, and a t-test does not. We will see later that this assumption yields different distributions for the test statistics under the null hypothesis.

It is known that the mean $\bar{x}$ is distributed as $N(\mu, \sigma/\sqrt{n})$, if x is normally distributed (or n is large). We can standardize the test statistic, as follows:

$$z = \frac{\bar{x} - \mu}{\sigma/\sqrt{n}},$$

where n is the number of observations in the sample. This yields the statistic that is typically used in the z-test to check a hypothesis about the mean of a population.

We now return to our example of the monthly rent for one-bedroom apartments to illustrate these concepts. Suppose that the distribution of rents in Manhattan is normal with a mean of $1,800 and a known standard deviation of $\sigma = 200$. This is a left-tail test, and we can determine the critical value as the value of x that yields $F(x) = \alpha$, where $F(x)$ is distributed as $N(\mu, \sigma/\sqrt{n})$. We can use the MATLAB inverse CDF function for the normal distribution to find the critical value as shown below. The critical region is

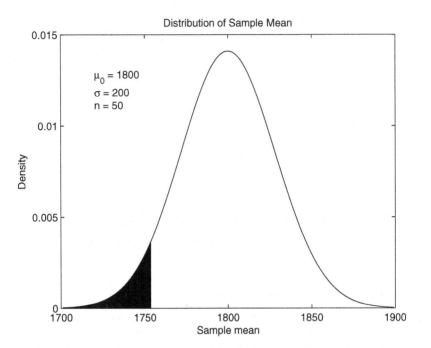

FIGURE 5.1

This shows the critical region for our hypothesis test that the monthly rent for one-bedroom apartments in Washington, D.C., is less on average than the monthly rent for similar apartments in Manhattan. The alternative hypothesis in this case corresponds to a left-tail test, so the critical region is in the lower part of the distribution.

illustrated in Figure 5.1. Note that we show the non-standardized version of the test statistic in the plot to make it easier to interpret.

The MATLAB code to get the critical value in terms of x is shown here:

```
mu0 = 1800;
sig = 200;
alpha = 0.05;
n = 50;
cv = norminv(alpha, mu0, sig/sqrt(n));
```

We get a critical value of $1,753.48. If our observed value of the test statistic is less than this number, then we can reject the null hypothesis in favor of the alternative hypothesis at the 5% level. Recall that the observed value was $1,750, which is less than the critical value. In other words, it falls within the critical region. Therefore, we have evidence that the monthly rent for one-bedroom apartments in Washington, D.C., is less than the cost in Manhattan.

We could also calculate the *p*-value using the CDF function:

```
pv = normcdf(1750, mu0, sig/sqrt(n));
```

```
% This result is shown in the window.
pv = 0.0385
```

Because the observed test statistic fell within the critical region, we should expect the *p*-value to be less than 0.05. It is, and we reject the null hypothesis.

The logic of the *t*-test works similarly and also assumes normality, but the test statistic is slightly different because we do not know the standard deviation. The test statistic for the *t*-test is

$$t = \frac{\bar{x} - \mu_0}{s/\sqrt{n}},$$

where *s* is the standard deviation calculated from the random sample. This test statistic has the Student's *t* distribution with $n - 1$ degrees of freedom. We would find the critical value (or region) using the Student's *t* distribution (**tinv** in the Statistics Toolbox) instead of the normal distribution.

```
% Let's assume that we do not know the population
% standard deviation. In this case, we have to use
% the Student's t distribution for the test statistic.
% First, find the degrees of freedom.
df = n - 1;

% Find the critical value.
cv_t = tinv(alpha, df)

% This critical value is returned to the window.
cv_t = -1.6766
```

The Student's *t* distribution is centered at zero, and it is parameterized by the degrees of freedom. If the observed value of the test statistic is less than the critical value (since this is a left-tail test), then we have evidence to reject H_0.

We have to find the observed value of the test statistic, as shown with the following steps.

```
% We estimated the standard deviation from the data
% and it had a value of 175.
sig_hat = 175;
t_obs = (1750 - 1800)/(sig_hat/sqrt(n))

% This value is observed.
t_obs = -2.0203
```

This is less than the critical value, and we reject the null hypothesis in favor of the alternative. In other words, we have evidence that the monthly rent in Washington is less than the rent in Manhattan. We can also get the *p*-value.

```
% Now, find the observed p-value.
pv = tcdf(t_obs,df)

% This is the observed p-value
pv = 0.0244
```

Fortunately, MATLAB has functions in the Statistics Toolbox that handle the details of the *z*-test and the *t*-test for us. The function

```
ztest(x,mu0,sigma,'alpha',val,'tail',type)
```

performs a *z*-test of the hypothesis that the data in the vector **x** come from a normal distribution with mean **mu0** and known standard deviation **sigma**. The arguments **'alpha'** and **'tail'** are optional. If they are left out, then the default is to conduct a two-tail test at the $\alpha = 0.05$ level. Options for the **'tail'** argument are **'right'**, **'left'**, and **'both'**.

The **ztest** function will return a logical value of 0, indicating that the null hypothesis cannot be rejected at the **alpha** level of significance. A value of 1 indicates that the null hypothesis can be rejected.

The function **ttest** has a similar syntax:

```
ttest(x,mu,'alpha',val,'tail',type)
```

You can also obtain the *p*-value and the confidence interval for both types of tests by using output arguments, as shown here:

```
[h,pv,ci] = ttest(x,mu,'alpha',val,'tail',type)
```

where **h** is the logical indicator of 0 or 1; **pv** is the *p*-value; and **ci** is the confidence interval. We describe how to use these hypothesis tests in the next section.

TIP

The Student's *t* distribution approaches a $N(0, 1)$ distribution, as the sample size gets larger. Therefore, if *n* is large and the population σ is unknown, then we can use the *z*-test instead of the *t*-test.

5.2.2 Examples of Hypothesis Tests

Example of the z-test

We return to Fisher's iris data to illustrate the *z*-test in MATLAB, where we will use the sepal length variable for our analysis. It is always a good idea to do a quick check of the distribution assumptions before conducting any hypothesis test. So, we first construct a boxplot to see if the normality assumption is reasonable. This is shown in Figure 5.2. Looking at the shapes of the distributions, it appears that the sepal length of *Setosa* is approximately normal, while the others have slightly skewed shapes.

```
% First do a boxplot of the sepal length.
load iris

% This is the first variable - extract it.
setSL = setosa(:,1);
virSL = virginica(:,1);
verSL = versicolor(:,1);

% Do a notched boxplot.
boxplot([setSL,virSL,verSL],'notch','on',...
    'labels',{'Setosa','Virginica','Versicolor'})
ylabel('Sepal Length')
title('Boxplots of Sepal Length in Iris Data')
```

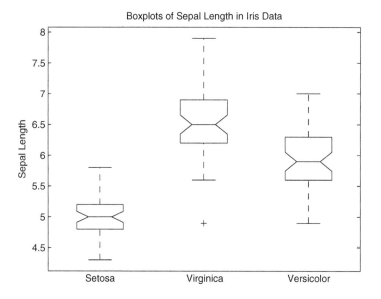

FIGURE 5.2

These are notched boxplots for the sepal length of Fisher's iris data. Note that the sepal length for Setosa appears to be normally distributed, while the others appear to be slightly skewed. The intervals given by the notches do not overlap across the three species, indicating that there is significant evidence at the 5% level that the medians are different. The IQR (or the height of the box) looks different between Setosa and the other two iris species.

As another visual check of how the data are distributed, we could create Q–Q plots or a normal probability plot, as we do in Figure 5.3. If the data fall on a straight line, then this indicates that they follow an approximate normal distribution. The **normplot** function in MATLAB will superimpose a line found using the first and third quartiles of the data to help with the assessment. Here is the MATLAB code to create the plot.

```
% As another check, do a normal probability plot.
normplot([setSL, virSL, verSL])
```

The observed sepal length for all three species fall on the lines for the most part, showing that the variables follow a normal distribution.

There is also an hypothesis test called *Lilliefors' composite goodness-of-fit test* [Lilliefors, 1967] that we can use to test the null hypothesis that the data come from a normal distribution. The MATLAB function for this test is called **lillietest**, and we apply it to the sepal lengths for each species of iris, as shown here.

```
% Now, do a Lilliefors' test for normality.
lillietest(setSL)
lillietest(virSL)
lillietest(verSL)
```

All of the results from **lillietest** are **0**, indicating that it failed to reject the null hypothesis that sepal length is normally distributed for each of the three iris species.

Based on these plots and the Lilliefors' test, it seems safe to assume that the *Setosa* sepal length is normally distributed. Let us further assume that the standard deviation is known, and it has a value of $\sigma = 0.3$. This means we can apply the z-test to test the hypothesis that

$$H_0: \quad \mu = \mu_0 = 5$$
$$H_1: \quad \mu \neq \mu_0 = 5$$

The value of $\mu_0 = 5$ was chosen for illustration only. The exact hypothesis test will depend on one's application.

This is a two-sided test because our alternative hypothesis is that the mean *Setosa* sepal length is not equal to $\mu_0 = 5$. This is the default for the **ztest** function, so we do not have to specify the tail region. Here is the function call.

```
[hyp, pv, ci] = ztest(setSL, 5, 0.3)
```

The results are:

```
hyp = 0
pv = 0.8875
ci = [4.9228, 5.0892]
```

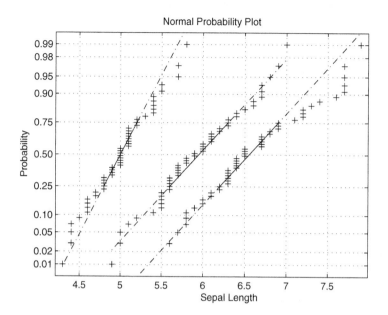

FIGURE 5.3

This normal probability plot for the sepal length of Fisher's iris data indicates that the normality assumption is a reasonable one because the observations mostly fall on the lines.

The logical output of **hyp** = 0 indicates a failure to reject the null hypothesis at the default level of 5%. Likewise, the high p-value of 0.8875 provides the same conclusion.

In the case of a two-tail test, we can also use the confidence interval to check our hypothesis. The 95% confidence interval from the **ztest** contains the population mean of $\mu_0 = 5$, which is consistent with the results from hypothesis testing.

Let's look at another example of the z-test using the sepal length of the iris *Versicolor* species. We want to test the following hypotheses:

$$H_0: \qquad \mu = \mu_0 = 5.7$$
$$H_1: \qquad \mu > \mu_0 = 5.7$$

Assume we have evidence from previous experiments that $\sigma = 0.6$. This is a right-tail test, as indicated in this call to the **ztest** function.

```
% An additional example with Versicolor
% sepal length and the alternative hypothesis that
% the mean is greater than 5.7.
[hyp,pv,ci] = ztest(verSL,5.7,0.6,'tail','right')
```

```
% These are the results from the z-test.
hyp = 1
pv = 0.0027
ci = [5.7964 , Inf)
```

A value of **hyp = 1** means that we have evidence to reject the null hypothesis that the mean sepal length of iris *Versicolor* is equal to 5.7. The p-value of 0.0027 indicates the same conclusion.

Example of the t-test

We turn to a new data set to illustrate the **ttest** function for a paired data set. These data were downloaded from the DASL website at this link:

http://lib.stat.cmu.edu/DASL/Datafiles/FishPrices.html

There are 14 observations for the prices in cents per pound for several species of fish in 1970 and 1980. We saved the data in a file called **fish.mat**. This file can be downloaded from the website for the book, along with the other data sets mentioned in Chapter 1.

Our first task is to gain some insight into how the data are distributed, and we construct a normal probability plot, as we did in the previous example.

```
% Load fish prices data.
load fish

% Construct a normal probability plot to test
% the normality assumption.
normplot([fish70, fish80])
```

The plot is shown in Figure 5.4, and we observe a slight deviation from the straight lines through the first and third quartiles of the samples. Our sample size is small, and deviations like this are not unreasonable, even if the data are normally distributed. We performed a Lilliefors' test for the prices in both years, and the **lillietest** function returned 0 in both cases, which means that we do not have evidence to reject the null hypothesis. This could be an indication that the underlying population is indeed normal, or it might mean that the sample size is too small to detect significant evidence for a departure from normality.

We can see from the normal probability plot in Figure 5.4 that the mean fish price seems to have shifted between 1970 and 1980. First, let us find the average fish price in 1970 and then test the null hypothesis that the mean price of fish in 1980 is the same, using a one-sample t-test.

```
% Find the mean of the fish prices in 1970.
mu70 = mean(fish70);
```

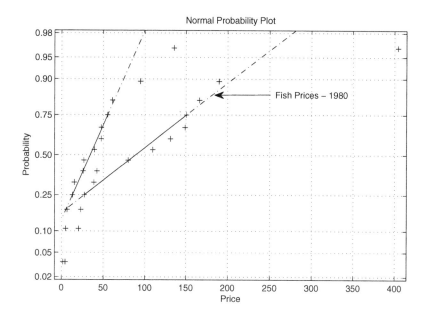

FIGURE 5.4
Here we have a normal probability plot for fish prices in 1970 and 1980. There is a slight deviation from the line, but a Lilliefors' test fails to reject the null hypothesis that the data are normally distributed. So, our normal assumption for these data seems to be all right. We also see a shift in the average fish price between 1970 and 1980.

```
% Do a t-test to see if the fish prices
% in 1980 have the same mean as in 1970.
[hyp, pv, ci] = ttest(fish80,mu70)
```

The output from the test is shown here.

```
hyp = 1
pv = 0.0296
ci = [49.0664 , 170.5336]
```

We see from these results that there is significant evidence at the 5% level that the average fish price in 1980 is different from the average price in 1970. In particular, the confidence interval does not contain the mean of the 1970 prices, which was 41 cents per pound.

Example of a paired t-test

The **fish** data are matched or paired because we have measurements for the price of a particular type of fish in 1970 and 1980. So, we can also perform a *paired t-test*. This test assumes that the difference between matched observations from two samples $x_a - x_b$ are a random sample from a normal distribution with mean zero and unknown variance. Thus, x_a is a unit in one sample, and x_b is the corresponding or matching unit from another sample. Here is the result of this test applied to the fish data, where we are testing the following hypotheses:

$$H_0: \quad \mu = \mu_{fish70} - \mu_{fish80} = 0$$
$$H_1: \quad \mu \neq 0 .$$

Here is the MATLAB code for a paired t-test.

```
% Ask for the paired t-test.
[hyp, pv, ci] = ttest(fish70, fish80)

% This is displayed in the window.
hyp = 1
pv = 0.0027
ci = [-108.7942 , -28.6058]
```

The result of **hyp = 1** tells us that the null hypothesis is rejected at the default 5% level of significance. In other words, it rejects the hypothesis that the average fish price in 1970 is the same as in 1980. We get the same result from the confidence interval because it does not contain zero. Furthermore, both ends of the confidence interval are negative, indicating that prices increased.

Next, we will test the alternative hypothesis that the average price per pound (in cents) increased in 1980 relative to prices in 1970 to demonstrate how we can perform a one-tail test in MATLAB. This would be a right-tail test, and this is how we call the function:

```
% Ask for the t-test.
hyp = ttest(fish80,mu70,'tail','right')

% We get a value of 1, which indicates significance.
hyp = 1
```

The result from our hypothesis test is **hyp = 1**, which indicates that we have significant evidence at the 5% level that the prices increased, and we should reject the null hypothesis that the prices stayed the same.

Example of a Two-Sample t-Test

Our final example in this section shows how to perform a two-sample t-test of the null hypothesis that two data sets are each normally distributed independent random samples with the same mean and unknown variances. The hypotheses are:

$$H_0: \qquad \mu_a = \mu_b$$
$$H_1: \qquad \mu_a \neq \mu_b$$

We can perform this test in MATLAB using two options—equal or unequal variances. Referring back to the boxplots in Figure 5.2, we see that the variances for *iris Setosa* and *iris Versicolor* do not appear to be equal. Thus, we use the unequal variance option when conducting the two-sample *t*-test, by specifying the **VarType** argument in the function call, as shown below.

```
[h,pv,ci] = ttest2(setSL, verSL,'VarType','unequal')
```

The following results are displayed in the command window.

```
h = 1
pv = 3.7467e-17
ci = -1.1057 ,-0.7543
```

Given the boxplots in Figure 5.2, it is not surprising that these results indicate that we should reject the null hypothesis at the 5% significance level in favor of the alternative hypothesis that the data are from populations with unequal means.

> *TIP*
>
> **The function for the chi-square goodness-of-fit test is called chi2gof, and it can be used for discrete or continuous probability distributions. It tests the null hypothesis that a sample comes from a specified distribution against the alternative that it does not.**

5.3 Confidence Intervals Using Bootstrap Resampling

We already discussed several MATLAB approaches for finding confidence intervals in Chapter 4. These options include using the **paramci** function or specifying an output argument from probability fitting and hypothesis testing functions. There is another option for finding confidence intervals, which is based on the bootstrap [Efron and Tibshirani, 1993]. We describe the basic bootstrap procedure first and then show how this can be used to obtain confidence intervals.

5.3.1 The Basic Bootstrap

The *bootstrap* is a method based on Monte Carlo simulations, and it is nonparametric in the sense that assumptions are not made about the distribution of the underlying population. Instead, we use the sample itself as an estimate of the population and also use it to estimate the sampling distribution of a statistic. Here are the basic steps for the bootstrap.

1. Calculate the statistic (mean, standard deviation, etc.) of interest from the original data set.
2. Obtain a random sample of size n, sampling *with replacement* from the original sample. This is called a **bootstrap sample**.
3. Calculate the same statistic in step 1 using the sample found in the second step. This is called a **bootstrap replicate statistic**.
4. Repeat steps 2 and 3, B times.
5. The replicate statistics form an estimate of the distribution for the statistic of interest and can be used to perform a hypothesis test, to estimate a standard error for the statistic, or to estimate a confidence interval [Efron and Tibshirani, 1993].

Note that the sampling is done *with replacement*, so an observation from the data might appear more than once in a bootstrap sample.

Another issue that must be addressed is to set a value for the number of bootstrap replicates B. If the statistic in step 1 is somewhat complicated, then a large value for B is appropriate. The choice also depends on the purpose for the bootstrap analysis (step 5). If you are going to estimate a standard error, then a value of B between 50 and 200 is usually sufficient. If your goal is to calculate a confidence interval, then B should be at least 1,000 [Efron and Tibshirani, 1993].

5.3.2 Examples

We will use the US temperature data to illustrate the bootstrap. Recall that we saved the data in a file called **UStemps.mat**, and we have to first load it into the workspace.

```
load UStemps     % Saved after importing - Chapter 1
```

This imports four variables or data objects into the workspace, two of which are **Lat** and **JanTemp**. Say we are interested in the relationship between the latitude of a city and its average minimum January temperature. We can plot the variables **Lat** and **JanTemp** in a scatter plot to get an idea of how they are related.

```
% Construct a scatter plot.
plot(Lat,JanTemp,'o')
xlabel('Latitude')
ylabel('Temperature ( \circ Fahrenheit )')
title('Average January Temperature - US Cities')
```

This plot is shown in Figure 5.5, and we see that the variables are negatively related. In other words, the average temperature decreases as the latitude increases. We can estimate the correlation between the two variables using the **corr** function.

```
cr = corr(Lat, JanTemp)

% We get this result.
cr = -0.8480
```

The correlation between the variables is close to -1, which indicates that there is a rather strong negative linear relationship.

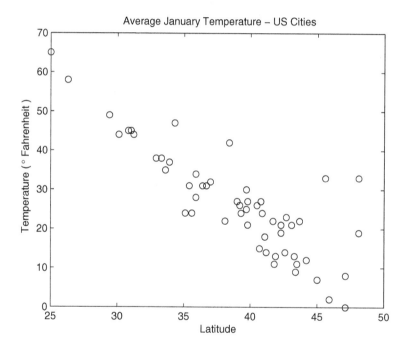

FIGURE 5.5

This is a scatter plot of the average minimum January temperature for 56 US cities. We see that there is a negative linear relationship between the two variables.

The **corr** function, which is part of the Statistics Toolbox, calculates a point estimate for the correlation coefficient. We can use the bootstrap approach to estimate a confidence interval for the correlation between latitude and temperature. The function is called **bootci**, and it is part of the Statistics Toolbox.

Example of Bootstrap Confidence Intervals

In the following example code, we ask for $B = 5,000$ bootstrap samples, because the correlation coefficient is a somewhat complicated statistic. The second input argument uses some unfamiliar MATLAB notation. The **@corr** input tells MATLAB what statistic is being calculated for each bootstrap sample. The **@** symbol is a pointer to a function, and the general use is **@functionname**. This argument is called the **bootfun** in the MATLAB documentation.

```
ci = bootci(5000, @corr, Lat, JanTemp)

% This is based on random bootstrap samples, so
% the resulting interval will not be exactly
% the same as the one we show here.
ci =

    -0.9299
    -0.6835
```

Thus, we have evidence that latitude and minimum January temperature are negatively correlated because both ends of the interval are negative. Please see Figure 5.6 for a histogram showing the distribution of the bootstrap replicated statistics and the location of the confidence interval.

The **bootci** function can produce other types of bootstrap confidence intervals. The default is to return the bias corrected and accelerated interval. Other options are the normal approximation with the bootstrap estimated standard error, the basic percentile method, and the studentized confidence interval [Efron and Tibshirani, 1993; Martinez and Martinez, 2007].

Here we show how to get the estimated confidence interval using the normal approximation and the bootstrap estimate of the standard error:

```
ci = bootci(5000, {@corr, Lat, JanTemp},...
     'type','normal')

% This is returned for the confidence interval.
ci =

    -0.9687
    -0.7332
```

Note that we had to enclose the information about our function **corr** in curly brackets, because we have additional input arguments that are not related to the **bootfun**. We see that the normal approximation produced a narrower interval than the fully nonparametric interval obtained previously. This is because it is based on some assumptions about the underlying distribution.

Example of Bootstrap Replicates

We conclude this section by showing how we can get other statistics based on the bootstrap samples. Recall that these are called the *bootstrap replicates* we calculate in step 3 of our procedure, and together they comprise an estimate of the distribution of our statistic. The function **bootstrp** will return the bootstrap replicates for a specified statistic. We can use these to visualize their distribution in a histogram, as shown in Figure 5.6 for the correlation coefficient between **Lat** and **JanTemp**.

```
% Get the bootstrap replicates.
bootrep = bootstrp(5000,@corr,Lat,JanTemp);

% Show their distribution in a histogram.
hist(bootrep,25)
xlabel('Estimated Correlation Coefficient')
ylabel('Frequency')
title('Histogram of Bootstrap Replicates')
```

We could also obtain an estimate of the standard error for the correlation statistic, as shown here.

```
se = std(bootrep)

% This result is shown in the window.
se = 0.0606
```

> **TIP**
>
> The standard error from the bootstrap can be used in hypothesis testing and to get confidence intervals.

5.4 Analysis of Variance

In Section 5.2, we showed how to test the hypothesis that the means of two groups are different. *Analysis of variance* or ANOVA is a methodology that enables us to compare the means of different groups, so it might be more informative to call it the **analysis of variation about means** [Bhattacharyya and Johnson, 1977]. There are many different types of ANOVA that can be performed in MATLAB (see Table 5.3), but we discuss only one of them in detail. This is called the **one-way analysis of variance**, and we can think of this as a generalization of the *t*-test we saw previously.

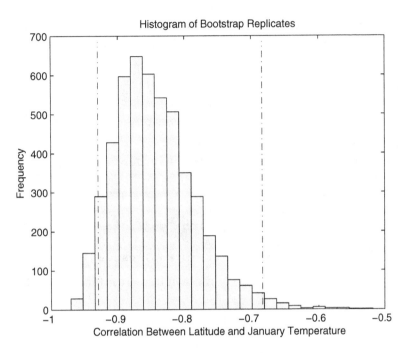

FIGURE 5.6
This shows the sampling distribution of the correlation coefficient statistic based on the bootstrap replicates. The bootstrap bias corrected and accelerated confidence interval is shown by the vertical lines. Note that the distribution looks slightly skewed, indicating that the normal approximation for a confidence interval might not be appropriate.

5.4.1 One-Way ANOVA

Suppose that we have k populations (or groups) that we are studying, and we take an independent random sample from each one. The statistical model that forms the basis for ANOVA is given as

$$x_{ij} = \mu_i + \varepsilon_{ij},$$

where x_{ij} is the j-th sample unit in the i-th group. The mean of the i-th group is μ_i. The error terms ε_{ij} are independent normally distributed random variables with a mean of zero and variance σ^2. The indexes take on values $i = 1, ..., k$ and $j = 1, ..., n_i$.

Our objective is to use the differences in the sample means to decide whether any observed differences are due to sampling variation or because the group means are different. It might help to see an example, where we have two sets of three groups or populations. The three samples in the left panel of Figure 5.7 appear to be grouped more about their own means, rather

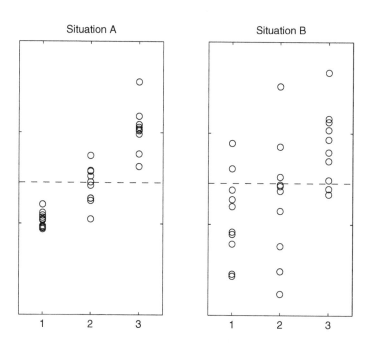

FIGURE 5.7
The three samples on the left panel appear to be tightly clustered around their individual group means instead of the overall mean of the data (dashed line). The panel on the right side shows three data sets that exhibit more variation within the groups.

than the overall mean (indicated by the dashed line). We can contrast this with the data shown on the right, where the individual samples are more dispersed. Based on the scatter plots, we would probably be inclined to say that there is evidence that the population means in situation A are different. Whereas, this does not seem likely in situation B. As we will see shortly, our test statistic will be based on these variations about the means.

We can express this as a hypothesis test, where we have the hypotheses

$$H_0: \mu_1 = \mu_2 = \ldots = \mu_k$$
$$H_1: \mu_i \neq \mu_j \qquad \text{for at least one pair}$$

The next step is to come up with a test statistic that will help us make our decision regarding this hypothesis. Looking at the data in Figure 5.7, it makes sense to have a test statistic that compares the variations in the sample from a single group to the variations among the groups.

Recall that the overall mean is denoted by $\bar{x}$. We will represent the mean for the i-th sample as $\bar{x}_i$. The **total sum of squares** measures the dispersion of all the data around the overall mean and is given by

$$\text{Total } SS = \sum_i \sum_j (x_{ij} - \bar{x})^2 .$$

We can break this up into two pieces representing variation within groups and between groups, as shown here

$$\sum_i \sum_j (x_{ij} - \bar{x})^2 = \sum_i \sum_j (x_{ij} - \bar{x}_i)^2 + \sum_i n_i (\bar{x}_i - \bar{x})^2$$

$$= SSE + SS_T$$

The first term on the right side measures the variation about the mean within groups, and it is called the **error sum of squares** (*SSE*). It is also sometimes called the **residual sum of squares**. The second term is the **treatment sum of squares** SS_T, and it is the variation of the group means with respect to the overall mean $\bar{x}$. The word *treatment* is used because ANOVA was originally developed for experimental design, where one is interested in comparing the effect of different treatments on an outcome variable. The examples will show that MATLAB refers to this as the sum-of-squares for the columns.

TIP

A *factor* is a condition of the experiment that can be controlled by the analyst. A specific combination of factor levels is a *treatment*.

We are now ready to develop a test statistic that utilizes the two types of variation on the right-hand side—the *SSE* and the SS_T. If there is truly no difference in the group means, then it is likely that *SSE* and SS_T would be approximately equal. If they are very different, then we have evidence against our null hypothesis. The test statistic is the F statistic given by

$$F = \frac{SS_T/(k-1)}{SSE/(n-k)},$$

where $k-1$ and $n-k$ are the degrees of freedom.

If the variation within groups (*SSE*) is small compared to the variation between groups (SS_T), then we would expect to observe a large value of the F statistic. This is the situation illustrated in the left panel of Figure 5.7. On the other hand, if the variation within the groups is large compared to the between-group variation, then the observed F statistic will be small.

The assumptions for the ANOVA *F*-test are:

- The samples from the *k* populations or groups are random and are independent from each other.
- The populations are normally distributed and have the same variance.

The *F* test statistic still works well when the assumption of normality and equal variances are violated [Bhattacharyya and Johnson, 1977].

5.4.2 ANOVA Example

We illustrate one-way ANOVA using the Fisher's iris data. In this case, the different populations or groups are given by the species. We use one-way ANOVA to test the null hypothesis that the group means for a specific iris characteristic are equal.

We first have to load the data into MATLAB and then create a matrix where each column has the observed characteristic for one of the groups.

```
load iris
% This gives us three separate data objects in
% the workspace — one for each species.
% We need to put one of the characteristics
% from all three groups into one matrix.
% Let's look at sepal width ...
sepW = [setosa(:,2),virginica(:,2),versicolor(:,2)];
```

We can perform the basic one-way ANOVA using the **anova1** function. This expects a matrix with each column holding the data for a group.

```
% Perform a one-way ANOVA.
[pval, anova_tbl, stats] = anova1(sepW);
```

This returns the following *p*-value in the command window:

```
% Type this to display the p-value.
pval

% This is displayed.
pval = 4.4920e-17
```

In addition, two figure windows are opened, as shown in Figure 5.8 for our example. One window contains side-by-side notched boxplots for the groups. This provides several things. First, it allows us to assess whether our assumptions of normality and equal variance are reasonable, which they seem to be for this example. Second, the notched boxplots give us a visual way to test whether the *medians* are different. If the intervals given by the notches do not overlap, then we have evidence that the *medians* are different.

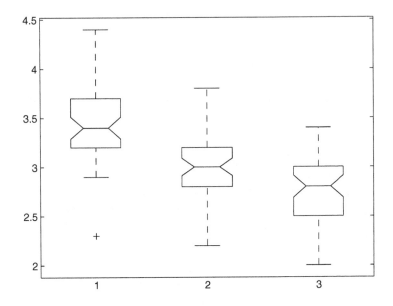

FIGURE 5.8
This is the graphical output from **anova1**. *The sepal width was used for this analysis, and we see a notched boxplot of the three groups. There appears to be significant evidence that the* **medians** *between group 1 (Setosa) and the others are different because the notched intervals do not overlap. However, it is not too clear with groups 2 (Virginica) and 3 (Versicolor). The results of the ANOVA test are shown in the table. The second column is the sum of squares (SS); the third column is the degrees of freedom; and the fourth column is the mean of squares (SS/df). The observed value of the F-statistic and the corresponding p-value are also shown. The p-value is very small, so we have evidence that at least one pair of group means is significantly different.*

The second window that is returned by the **anova1** function is the ANOVA table. The standard ANOVA table divides the variability in the data due to the different sources we have already discussed—variability between groups and variability within groups. The ANOVA table has six columns:

- Source of the variability.
- The sum of squares (SS) due to each source.
- The degrees of freedom (df) associated with each source.
- The mean squares (MS) for each source, which is the ratio SS/df.
- The observed value of the F statistic, which is the ratio of the mean squares.
- The p-value, which is obtained using the CDF of the F statistic.

The alternative hypothesis in a one-way ANOVA is that at least one pair of group means is significantly different. However, what pair is different? We can use the *multiple comparison test* in the Statistics Toolbox to determine which pair is different. Note that it requires the **stats** output from **anova1**.

```
comp = multcompare(stats)

% This is the results of the comparison test.
comp =

    1.0000    2.0000    0.2948    0.4540    0.6132
    1.0000    3.0000    0.4988    0.6580    0.8172
    2.0000    3.0000    0.0448    0.2040    0.3632
```

The first two columns of this output represent the group numbers. Thus, the first row is comparing groups 1 and 2, and the last row compares groups 2 and 3. The third and fifth columns are the end points of a 95% confidence interval for the *difference of the group means*, and the fourth column is the estimated difference. Thus, the difference in the means of groups 1 and 3 is **0.6580**, and the confidence interval for the difference is given by

$$[0.4988 , 0.8172].$$

The interval does not contain zero, so we can conclude that this pair of means is significantly different.

We also get some other helpful graphical output from the **multcompare** function. MATLAB automatically opens a window, similar to what is shown in Figure 5.9. There will be a horizontal line for each group, where the line is a graphical representation of the estimated mean and a 95% comparison interval for the mean. You can click on each of the lines in the plot, and MATLAB will display the result of the multiple comparison test. Groups with significantly different means will be shown in red.

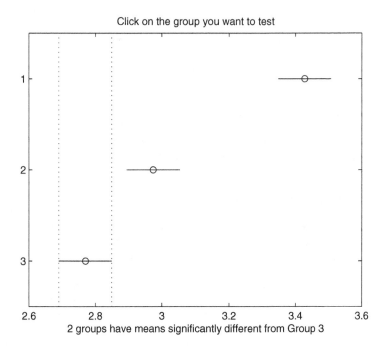

FIGURE 5.9

This is the graphical output from the **multcompare** *function. There is one line per group. The circle represents the estimated mean for the group, and the line is the 95% comparison interval for the estimated mean. You can click on each of the lines to find out what other groups are significantly different from the one selected. If we do that for this example, then we find that clicking on each one in turn shows that the remaining two pairs are significantly different.*

The **multcompare** function can be applied to other estimates—not just the means, as shown here. The graphical output from **multcompare** will represent the appropriate estimate. See **help** on **multcompare** for more information.

5.5 Summary and Further Reading

In this chapter, we provided an overview of hypothesis testing concepts and the MATLAB functions available for some of the commonly used tests. All of these functions are in the Statistics Toolbox; a partial list is given in Table 5.1. The online documentation for the Statistics Toolbox is an excellent resource for more information and examples on hypothesis testing. There is an entire chapter on hypothesis testing. The documentation can be accessed from the MATLAB desktop ribbon interface or at the link given here:

http://www.mathworks.com/help/stats/index.html.

It is difficult to recommend specific books on hypothesis testing, because many have been published. All introductory statistics books have chapters on most of the hypothesis testing concepts discussed in this chapter and can serve as resources for further study. See the book by Martinez and Martinez [2007] for more information on how simulations can be used for hypothesis testing and for the construction of confidence intervals using MATLAB.

TABLE 5.1

Some Hypothesis Tests in the Statistics Toolbox

Name of Test	Function Name	Null Hypothesis	Alternative Hypothesis
Ansari-Bradley	`ansaribradley`	Same distribution and variances	Different variances
Goodness-of-fit	`chi2gof`	Specified distribution	It is not from specified distribution
Friedman's Test	`friedman`	Column effects same	Not all the same
Jarque-Bera	`jbtest`	Normal, unknown mean and variance	It does not come from a normal distribution
Kruskal-Wallis	`kruskalwallis`	Multiple samples drawn from same population	They are not drawn from the same population
Kolmogorov-Smirnov	`kstest`	Sample comes from a distribution with specified parameters	It does not come from that distribution
Lilliefors	`lillietest`	Sample comes from a normal distribution	It does not come from a normal
Wilcoxon	`ranksum`	Same distribution with equal medians	Do not have equal medians

In this chapter, we also described how to construct confidence intervals using the bootstrap approach. There is a very readable book by Efron and Tibshirani [1993] on the bootstrap. An excellent review paper by the same authors is available via open access [Efron and Tibshirani, 1986]; the link is given in the references.

Table 5.2 lists the MATLAB functions for performing bootstrap and related tasks. For more information and examples on the bootstrap, refer to the *Resampling Techniques* section in the *Probability Distributions* chapter of the Statistics Toolbox online documentation.

TABLE 5.2

Bootstrap Functions

bootstrp	Calculate bootstrap statistics
bootci	Obtain a bootstrap confidence interval
datasample	Random sampling—with or without replacement
randsample	Get a random sample—with or without replacement

In the last section of this chapter, we described how to perform the basic one-way ANOVA. There are other types of ANOVA, and we discuss them briefly here. First, there is a two-way ANOVA that uses two categories to define the treatment levels. MATLAB provides a function called **anova2** for this situation. There is also a function called **anovan** that extends this idea to groups given by multiple factors. The **anovan** function also allows one to fit a model with random effects, where the levels are considered to be a random selection from a larger population.

The MATLAB functions for conducting ANOVA and related tasks are summarized in Table 5.3. For more information and examples on ANOVA, please see the *ANOVA* section in the *Regression and ANOVA* chapter of the Statistics Toolbox online documentation.

ANOVA is another topic that is usually covered in introductory statistics texts. So, there are a lot of options for further reading. There are also some books just on the analysis of variance, such as *Introduction to Analysis of Variance: Design, Analysis & Interpretation* [Turner and Thayer, 2001], *The Analysis of Variance: Fixed, Random and Mixed Models* [Sahai and Ageel, 2000], and *Primer of Applied Regression & Analysis of Variance* [Glantz and Slinker, 2001].

TABLE 5.3

Analysis of Variance Functions

anova1	One-way analysis of variance
anova2	Two-way analysis of variance
anovan	*N*-way ANOVA includes random effects
manova1	One-way Multivariate ANOVA
multcompare	Multiple comparison test

Chapter 6

Model-Building with Regression Analysis

Essentially, all models are wrong, but some are useful.
George E. P. Box (1919–2013)

This quote comes from the book on model-building and response surfaces by Box and Draper [1987]. It is quoted often because it provides useful insights about the nature of building and using models. In many data analysis applications, we are interested in creating models to represent some observed phenomenon. Statisticians collect data and use these to create the models that hopefully capture the phenomenon. Models are just a representation of some real process and are likely based on our own ideas of reality. Thus, they are typically wrong and will have some error associated with them.

However, all is not lost. We can still find a use for models, and we can use statistical methods to help us build the models and to assess the error. In statistics, the process of estimating or fitting models based on data and assessing the results is called *regression analysis*. In this chapter, we present some basic ways of estimating or building parametric models in MATLAB®. This includes fitting polynomials, least squares linear regression, and using the Basic Fitting GUI, which is available via a Figure window.

6.1 Introduction to Linear Models

In this section, we will be covering some basic concepts associated with linear models. First, we discuss the univariate case, where we have one *predictor variable* X and a *response variable* Y. A predictor variable is also known as an *independent* or *explanatory variable*, and the response is often called the *dependent variable*. We are interested in modeling the relationship between the two variables.

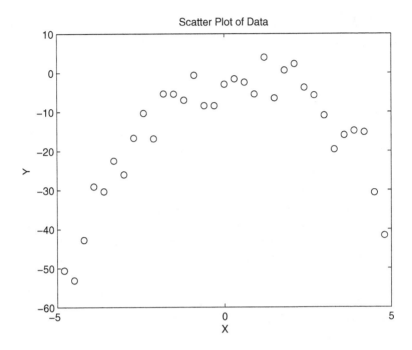

FIGURE 6.1
We generated some data that we will use in the first example. We want to estimate a rela-
tionship between the response or dependent variable Y and the predictor or independent
variable X. Visualizing the data in a scatter plot or other suitable plot (see Chapter 7)
should always be the first step in model building. The scatter plot shown here indicates that
a quadratic relationship between the variables might be reasonable.

6.1.1 Specifying Models

The first step in model-building is to collect data that measure the response
and predictor variables and plot these. The i-th observed value for a predictor
is denoted by x_i, and the corresponding response is represented by y_i, with
$i = 1, ..., n$. The next step is to propose a model for the relationship, based
on these data and knowledge of the subject area.

It is always a good idea to construct a scatter plot to help us come up with
a suitable model, as we show in Figure 6.1. The data shown in this plot will
be used in the first example given in this chapter. We start by giving some
examples of parametric models, such as a *linear first-order model*. This is given
by

$$Y = \beta_0 + \beta_1 X + \varepsilon,$$

where β_0 and β_1 are parameters that we estimate using the observed data, and ε represents the error. The word *linear* refers to the linearity of the β_j parameters and not to the fact that we might model the relationship of Y to X with a straight line. The **order** or **degree** of the model refers to the highest power of the independent variable X and determines the shape of the fitted curve. Thus, our straight line relationship given above is of first order. As another example, a *linear univariate second-order model* would be specified as

$$Y = \beta_0 + \beta_1 X + \beta_2 X^2 + \varepsilon.$$

We could extend this to multiple predictors $X_j, j = 1, ..., p$. An example of this type of model—with first-order terms only—is

$$Y = \beta_0 + \beta_1 X_1 + ... + \beta_p X_p + \varepsilon.$$

We can also have interaction terms (e.g., $X_1 X_2$) or other functions of the independent variables in our model, as shown here

$$Y = \beta_0 + \beta_1 X_1 + \beta_2 X_2 + \beta_3 X_1 X_2 + \beta_4 X_1^2 + \beta_5 \frac{1}{X_3} + \varepsilon.$$

We will only look at linear models in this primer, but we can have models that are *not* linear in the coefficients β_j. The following models are *not* linear:

$$Y = \frac{\beta_0 + \beta_1 X_1}{1 + \beta_2 X_1^2} + \varepsilon$$

$$Y = \beta_0 + \beta_1 X_1^{\beta_2} + \varepsilon$$

$$Y = \beta_0 + \sqrt{\beta_1} X_1 + \varepsilon.$$

Once we have the data, these models can be represented in matrix form as

$$\mathbf{Y} = \mathbf{X}\beta + \varepsilon,$$

where β is a column vector of parameters that we need to estimate from the data, and the $n \times 1$ vector ε denotes the errors. The n observed responses are in the vector $\mathbf{Y}$, and each row of the matrix $\mathbf{X}$ contains values for the observed predictors, according to the postulated model.

6.1.2 The Least Squares Approach for Estimation

To fit the model based on the data, we need to estimate the parameters β_j using the least squares approach. We get the least squares fit to the model by finding the values of the parameters that minimize the sum of the squared errors, which are shown here for a first-order model:

$$RSE = \sum_{i=1}^{n} \varepsilon_i^2 = \sum_{i=1}^{n} [y_i - (\beta_0 + \beta_1 x_i)]^2,$$

where *RSE* denotes the **residual squared error**. The approach uses concepts from calculus to obtain the values of β_j that minimize the *RSE*. We omit the details here since they can be found in most statistics texts.

The least squares solution for the parameters β can be found using this matrix expression

$$\hat{\beta} = (\mathbf{X}^T \mathbf{X})^{-1} \mathbf{X}^T \mathbf{Y}.$$

The least squares assumptions are listed here.

- The errors ε_i are independent.
- The errors ε_i are normally distributed as $N(0, \sigma)$. This means they have a common variance and a mean of zero.

We can use the least squares approach to estimate our model parameters, whether or not these assumptions are valid. However, we need them to hold when we want to conduct hypothesis tests on the significance of the terms in our model or to construct confidence intervals for the estimated coefficients. We discuss this in more detail later in this chapter.

The method described above is used for **parametric regression analysis**. We can use it to estimate the parameters β_j in our models because they are parametric models. An alternative approach called **nonparametric regression** is used when we have an unknown model or we do not want to assume linearity in the coefficients. Instead, we assume that there is a smooth relationship between Y and X. An example for such a model with one predictor is shown here:

$$Y = f(X) + \varepsilon.$$

A detailed discussion of nonparametric regression is beyond the scope of this book. However, we briefly list some MATLAB functions for nonparametric regression at the end of this chapter. We will concentrate only on linear parametric regression in the remainder of the chapter.

6.1.3 Assessing Model Estimates

We now provide information on some of the statistics used to evaluate the fitted models we obtained from our data. These include the coefficient of determination, 95% confidence intervals for the coefficients, and the *F*-statistic. We will highlight these when we describe the various MATLAB functions for regression analysis.

First, we explain a statistic that allows us to assess the overall fit of our model. This is called the *coefficient of determination* or the *R-squared*. The R-squared provides a measure of how well our model fits the data, and it can be derived as a decomposition of the variability in our *y* values. This is similar to the ANOVA approach discussed in Chapter 5.

The variability (sum of squares or SS) in our *y* values can be decomposed as follows:

Total SS of *y* = SS from Regression + Residual SS.

The total sum of squares on the left side is given by

$$\text{Total SS of } y \;=\; \sum (y_i - \bar{y})^2 .$$

The sum of squares from the regression is

$$\text{SS from Regression} \;=\; \sum (\hat{y}_i - \bar{y})^2 ,$$

where $\hat{y}_i$ is the response at x_i based on the estimated model. This is considered to be the sum of squares explained by the linear regression. Finally, the residual sum of squares is

$$\text{Residual SS} \;=\; \sum (y_i - \hat{y}_i)^2 .$$

This is the variability that is unexplained by the linear regression. The coefficient of determination r^2 is given by

$$r^2 \;=\; \frac{\text{SS from regression}}{\text{Total SS of } y} \;=\; 1 - \frac{\text{Residual SS}}{\text{Total SS of } y} .$$

When ordinary least squares is used to build the model, then $0 \le r^2 \le 1$. A value close to one indicates that the fitted model is accounting for more variability in the response variable *y*, as compared to the residual variability. That is, the sum of squares from the regression should be a major portion of the total sum of squares.

It can be shown that adding additional predictor variables and terms to the model will increase the r^2, but they might not be meaningful in terms of explaining the variability of y. There is an alternative statistic that accounts for this issue, and it is called the ***adjusted R-squared***. This applies a penalty for making the model more complex by adding terms to the model. Therefore, it will always be less than or equal to r^2. The adjusted R-squared is typically used to compare models as new terms are added to the model.

The R-squared is a way to assess the overall fit of the model by looking at what proportion of the variation in the observed responses can be explained by the fitted model. We can also assess the overall model by performing an hypothesis test based on the F-statistic. This is called the ***overall F-test for linear regression***. It tests the following null and alternative hypotheses:

$$H_0: \qquad \beta_1 = \beta_2 = \ldots = \beta_k = 0$$
$$H_1: \qquad \beta_j \neq 0 \quad \text{for at least one } j$$

The R-squared and the F-statistic are somewhat related to each other, in the following sense. If we have a strong relationship between our response and predictors as indicated by the R-squared, then the F-statistic will be large. However, if we add variables, and they do not produce large increases in the R-squared values for the model, then the F-statistic might actually decrease.

We see from the above null and alternative hypotheses that the F-test looks at all of the β coefficients except the constant. The F-statistic returns a result about the significance of at least one coefficient being nonzero, but it does not tell us which one is significant. For this, we can use a t-test applied to each of the coefficients. The hypotheses for a test on coefficient β_j are

$$H_0: \qquad \beta_j = 0$$
$$H_1: \qquad \beta_j \neq 0$$

Because this is a two-tailed test, we can perform an equivalent hypothesis test using confidence intervals. If we calculate a 95% confidence interval for a coefficient β_j and the interval does not contain zero, then we can conclude at the 5% level of significance that $\beta_j \neq 0$. In the context of model-building, this means that the corresponding term should be kept in the model.

6.2 Model-Building Functions in Base MATLAB®

The base MATLAB software provides functions for fitting linear models. These include functions for estimating polynomials, solving systems of equations, and least squares approaches.

We will be changing our notation for this section, because base MATLAB uses a slightly different notation than what we had in the previous section. The base software and documentation follows the notation and conventions from mathematics and linear algebra. For example, the models are usually stated in decreasing order of the terms, as shown here for the quadratic case

$$Y = \beta_2 X_1^2 + \beta_1 X_1 + \beta_0 .$$

Additionally, the matrix notation that is typically used in linear algebra for a system of equations is

$$\mathbf{A}\mathbf{x} = \mathbf{b} ,$$

where the interest is in solving for the unknown $\mathbf{x}$.

6.2.1 Fitting Polynomials

We first discuss estimating models with one predictor, where we will use a parameterization given by a polynomial. The main function for this purpose is called **polyfit**, and the basic syntax for the function is

```
p = polyfit(x,y,deg)
```

where **deg** is the order of the polynomial to be estimated. The output **p** is a vector of estimated coefficients β_j in *descending* order.

Our first example uses the data shown in Figure 6.1, which were obtained using the true relationship shown here:

$$y = -2x^2 + 2x + 3 .$$

We generated noisy data according to this relationship using the following MATLAB code.

```
% Generate predictor and response data using
% the true relationship and adding noise.
x = -4.8:0.3:4.8;
y = -2*x.^2 + 2*x + 3 + 5*randn(size(x));
```

Now, we use the **polyfit** function to estimate the parameters.

```
% Estimate the parameters.
p = polyfit(x,y,2)

p =

    -1.8009    1.9730    -0.5528
```

Keeping in mind that the estimated coefficients $\hat{\beta}_j$ are those accompanying the terms in descending order, we have

$$\hat{\beta}_2 = -1.8009$$
$$\hat{\beta}_1 = 1.9730$$
$$\hat{\beta}_0 = -0.5228.$$

The *hat* above the β denotes an estimate, and we see that our estimated model

$$\hat{y} = -1.8009x^2 + 1.9730x - 0.5228$$

is not too far from the true underlying relationship between x and y.

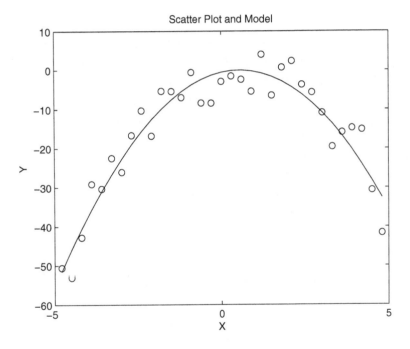

FIGURE 6.2
This shows the scatter plot of the data, along with the quadratic polynomial model that was estimated using **polyfit**.

MATLAB provides a function called **polyval** that uses the output from **polyfit** to evaluate the polynomial given by the vector **p** at given locations for x. We use this function to find estimated responses based on the fitted model. We then add that to the scatter plot of the data, as shown in Figure 6.2. Here is the MATLAB code to generate the estimated responses and the plot.

```
% Evaluate the polynomial and plot.
yhat = polyval(p,x);
plot(x,y,'o',x,yhat)
title('Scatter Plot and Model')
xlabel('X')
ylabel('Y')
```

Our next example builds a model for the relationship between the latitude and the average minimum January temperature of US cities. Recall that we have measurements of location and temperature for 56 cities. In our first model, the independent or predictor variable is the latitude, and the response or dependent variable is the temperature. First, we load the data into the workspace and create the scatter plot shown in Figure 6.3.

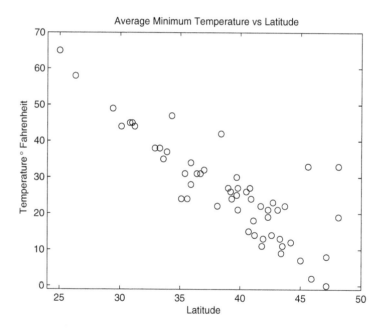

FIGURE 6.3
This is a scatter plot of January temperature versus latitude. A straight line seems to be a good model for the relationship.

```
load UStemps
% Create a scatter plot.
plot(Lat,JanTemp,'o')
```

A first-order model appears to be a reasonable one.

Next, we use **polyfit** to fit our model using the least squares approach, as shown below. Note that we are asking for an additional output argument that has information we will use shortly with the **polyval** function.

```
% Call the polyfit function with different outputs.
[ptemp,Stemp] = polyfit(Lat,JanTemp,1);
ptemp =

   -2.1096   108.7277
```

Our estimated coefficient $\hat{\beta}_1$ is equal to -2.1096. This is also known as the *slope* of the first-order model. It is negative, which makes sense from the scatter plot because temperature decreases as the latitude increases. The estimated model is

$$\hat{y} = -2.1096x + 108.7277.$$

The output variable **Stemp** is a MATLAB structure that can be passed to **polyval**. We display the **Stemp** object here.

```
display(Stemp)

Stemp =

       R: [2x2 double]
      df: 54
   normr: 52.5843
```

We can use this output object to generate statistics that help us understand our prediction error. These statistics are the standard deviation of the errors when predicting a future observation x based on the model, and we denote them by Δ. As stated previously, the usual assumption regarding the errors ε is that they are independently distributed as $N(0, \sigma)$. If this is the case, then $\hat{y} \pm \Delta$ will contain at least 50% of the predictions of future observations at a given value of x. We now show how to obtain Δ for the estimated model of US temperatures as a function of latitude.

```
% Generate some different X values.
xp = linspace(min(Lat),max(Lat),10);

% Get estimates using the model.
[ust_hat,delta] = polyval(ptemp,xp,Stemp);
plot(Lat,JanTemp,'o')
axis([24 50 -1 70])
xlabel('Latitude')
ylabel('Temperature \circ Fahrenheit')
title('Average Minimum Temperature vs Latitude')
hold on
```

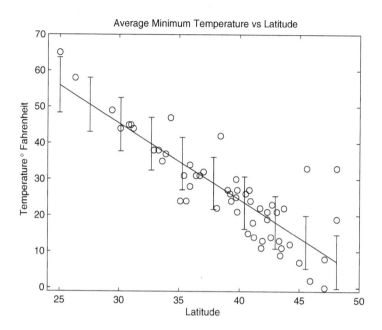

FIGURE 6.4

This is a scatter plot of the US temperature data with the first-order model superimposed on the plot. We also display bars that denote the standard deviation of the error in predicting a temperature for a given latitude. If the errors in the original data are normally distributed with mean zero and constant variance, then the bars represent an interval that will contain at least 50% of the predictions of future observations at the given values of x.

```
% This plots the data and adds error bars.
errorbar(xp, ust_hat, delta), hold off
```

The results of this code are shown in Figure 6.4, where we see that our model for the relationship between temperature and latitude seems plausible.

6.2.2 Using the Division Operators

MATLAB provides special division operators that can be used to estimate the parameters in our model. In this section, we describe these operators, along with their equivalent functional forms—**mldivide** and **mrdivide**.

Recall from the previous section that the matrix notation in linear algebra for a system of equations is usually denoted as

$$\mathbf{A}\mathbf{x} = \mathbf{b},$$

with the $m \times n$ matrix **A** being equivalent to our matrix **X** from before (see Section 6.1), and **b** containing the observed responses. In this matrix representation, we want to solve for **x**.

There are three situations we might encounter, depending on the structure of **A**. First, the matrix **A** could have the same number of rows and columns ($m = n$), in which case, we might find an exact solution. However, in statistical analysis, we typically have the situation, where $m > n$, and we have an over-determined system. This was the case with the US temperature data. If $m < n$, then we have fewer equations than unknowns to find, and we can obtain a solution with at most **m** nonzero components. This is called an under-determined system.

MATLAB's backslash or left-divide operator can be used to solve the above matrix equation, and it can handle all three cases we just described. The solution to the system of equations is denoted as

$$x = A \backslash b.$$

If **A** is a scalar, then the left-divide operator will perform element-wise division, which is the same as **A. \b**. Otherwise, it finds the least squares solution. There is also a function called **mldivide** that can be used instead of the left-divide operator. However, the more efficient approach is to use the backslash.

We return to our first example in this chapter, where we generated some data from a second-order polynomial, and we show how to find the solution using the left-divide operator. First, we have to create the matrix **A** (**X** using the original notation). This is typically called the *design matrix* in statistics.

```
% This example uses the data generated from a
% second-order polynomial. Create a design matrix
% 'A' using a quadratic model. 'x' is the solution
% vector, so we specify the observed predictors.
pred = x(:);
A = [pred.^2, pred, ones(size(pred))];
```

The response vector needs to be a column, so we use the colon notation to ensure that we have one.

```
% The response variable needs to be a column vector.
% We continue to use the alternative notation.
b = y(:);
```

Now, we find the solution using the left-divide operator.

```
x = A\b

x =
```

```
-1.8009
 1.9730
-0.5528
```

This is the same solution we obtained using the **polyfit** function.

There is also a right-divide operator, which is given by the forward slash or the usual division operation symbol. This solves the system of equations given by

$$xA = b.$$

Using the right-divide operation, the solution is given by

$$x = b/A.$$

This situation is not very common in practice. As with the left-divide, there is an alternative function called **mrdivide** that can be used instead of the slash.

<u>*TIP*</u>

There is a function in base MATLAB called **linsolve**. This will solve the system of equations given by $Ax = b$.

6.2.3 Ordinary Least Squares

The base MATLAB software has another function called **lscov** that will also find the least squares solution to a linear model. However, it has more options that data analysts might find useful. In particular, one can get other information from the function, such as the mean squared error (MSE) of the error term and the standard errors of the estimated regression coefficients.

We now return to the **UStemps** data to illustrate the use of **lscov** to find the estimates of our coefficients using ordinary least squares. The four data objects (**City**, **JanTemp**, **Lat**, **Long**) should still be in our workspace. This time, we will model the relationship of the minimum January temperature as a function of both latitude and longitude. From one of our previous examples (see Figure 6.3), we speculate that a linear relationship between temperature and latitude is a good one, but how is *longitude* related to temperature? We can look at a scatter plot of these two variables, as shown in Figure 6.5.

```
% Create a scatter plot of temperature and longitude.
plot(Long,JanTemp,'.')
xlabel('Longitude')
title('Minimum January Temperature')
ylabel('Temperature \circ Fahrenheit')
```

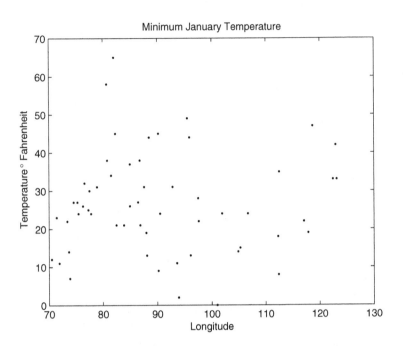

FIGURE 6.5

This is a scatter plot of the minimum January temperature versus longitude. We will model the relationship as a straight line (first-order), but it is apparent from the plot that this linear model might not be a good choice. We will explore this idea later.

From the plot, we see that a straight line might not be a good model for the relationship between January temperature and longitude. However, we will start with that one for illustration purposes, and an alternative one will be explored later in this chapter.

This type of model is called a ***multiple regression model*** because we have two or more predictor variables. The examples we have given so far had only one variable. The model we will use is first-order in each of the predictor variables, as shown here

$$Y = \beta_0 + \beta_1 X_{Lat} + \beta_2 X_{Long}.$$

X_{Lat} is the latitude, and X_{Long} is the longitude.

The following MATLAB code is used to set up the model and to estimate the coefficients. We first have to create the design matrix **A** according to our model. In our example, this matrix will contain a column of ones to represent the constant term in our model. Next, we use the **lscov** function, where we ask for the estimated coefficients, their standard errors, and the mean squared error.

```
% First step is to create the design matrix A.
% We will include a constant term and two
% first-order terms for latitude and longitude.
At = [ones(56,1), Lat, Long];
[x,stdx,mse] = lscov(At,JanTemp)
```

Here are the results.

```
x =

    98.6452
    -2.1636
     0.1340

stdx =

     8.3271
     0.1757
     0.0631

mse =

    48.0877
```

The standard errors can be used to create confidence intervals, as discussed in Chapter 5. Recall that our model uses the assumption that the coefficients have a common variance given by σ^2. The mean square error can be used as an estimate of the variance σ^2.

The MATLAB functions described so far implement an approach known as *ordinary least squares* to distinguish it from an alternative approach called *weighted least squares*. The **lscov** function will also perform weighted least squares. In this case, one supplies weights (e.g., inverse of the variance) for the observations in the function call, in order to reduce the influence of potential outliers on the model fit. The interested reader can refer to the MATLAB documentation on **lscov** for more information and examples for this type of model.

6.3 Functions in the Statistics Toolbox

The Statistics Toolbox provides additional functionality for building models, and we describe only the ones for basic linear parametric regression in this section. Other options for both parametric and nonparametric regression are provided in the last section of the chapter.

There are four main functions in the Statistics Toolbox for fitting linear regression models, and their availability depends on the MATLAB version being used. They are summarized in Table 6.1, with the second column of the table showing the basic syntax to call the function. As always, one should check the **help** and documentation for other options.

In the previous section, we saw an example of a design matrix for a model that has a constant term. We had to add a column of ones to represent this term in the model. The functions in the Statistics Toolbox have different requirements regarding the need for a constant term in a design matrix, and we indicate these in the last column of Table 6.1.

TABLE 6.1

Linear Regression Functions in the Statistics Toolbox

Function Name	Function Call (basic syntax)	Column of Ones Needed
regress	b = regress(y,X)	YES
regstats	stats = regstats(y,X,model)	NO
LinearModel.fit[a]	mdl = LinearModel.fit(y,X,model)	NO
fitlm	lm = fitlm(X,y,model)	NO

[a]The function **LinearModel.fit** is available in MATLAB R2012a or later, and the function **fitlm** was introduced in version R2013b. **LinearModel.fit** will be removed in later versions of MATLAB.

We will now be using the notation that is commonly employed in the statistics community in order to be consistent with the *Statistics Toolbox User's Guide* and related MATLAB documentation. Recall that in statistical analysis, the matrix formulation of the problem is given by

$$y = X\beta + \varepsilon,$$

where the design matrix is given by X; the errors are represented by ε; and the vector of responses are in the vector y. In this section, we will expand our regression analysis to include hypothesis testing with our model and the estimation of confidence intervals.

6.3.1 Using **regress** for Regression Analysis

One of the earliest regression functions available in the Statistics Toolbox is called **regress**. This function works for linear parametric regression models only. The complete function call for **regress** is

```
[b,bint,r,rint,stats] = regress(y,X,alpha)
```

The arguments for the function are described here.

- **b** is the vector of estimated coefficients, and **bint** contains confidence intervals for the estimated coefficients.
- **r** is a vector of residuals (or estimated errors), and **rint** has intervals that can be used to detect outliers.
- **stats** is a vector containing the R-squared (r^2) statistic, the F-statistic, the p-value of the F-statistic, and an estimate of the error variance σ^2.
- The default in **regress** is to produce 95% confidence intervals for the estimated coefficients and residuals. The **alpha** argument can be used to change this.

We will use the US temperature data for all of the examples in Section 6.3, and we will start with the same model as before, which is repeated here:

$$Y = \beta_0 + \beta_1 X_{Lat} + \beta_2 X_{Long}.$$

Our first step is to create the design matrix according to this model, but this time we will use the notation from statistics and call it **X**.

```
% We need our design matrix X.
% The regress function requires a column of ones.
X = [ones(56,1), Lat, Long];
```

Next, we call the **regress** function and ask for all of the output arguments.

```
[b,bint,r,rint,stats] = regress(JanTemp,X);
```

Here are the estimates for the coefficients and their 95% confidence intervals.

```
b =

    98.6452
    -2.1636
     0.1340

bint =
```

```
81.9432    115.3472
-2.5160     -1.8111
 0.0073      0.2606
```

We see that this matches the estimates we obtained in the previous section using **lscov**. Notice that the 95% confidence intervals for β_1 and β_2 do not contain zero, which implies that they are significant at the 5% ($\alpha = 0.05$) level and should remain in the model.

The **stats** vector contains some useful information. Let's look at the object we get from our example.

```
display(stats)

stats =

    0.7411    75.8753    0.0000    48.0877
```

The first element of **stats** is the coefficient of determination, so we have an $r^2 = 0.7411$, which is an indication that our model explains about 74% of the variability in the January temperature.

The second element of the **stats** vector is the F statistic, which is 75.8753. In this application, the F statistic is used to test the null hypothesis of a constant model (i.e., $\beta_{Lat} = \beta_{Long} = 0$) against the alternative model that at least one of these coefficients is nonzero. The p-value associated with the F statistic is essentially equal to zero, as seen when we extract it from **stats**.

```
stats(3)

ans =

    2.7915e-16
```

These results mean that our model is significant at the $\alpha = 0.05$ level or that at least one of our coefficients for **Lat** and **Long** is nonzero. The last element in **stats** is an estimate of the error variance, which in this example is 48.0877.

6.3.2 Using **regstats** for Regression Analysis

The **regstats** function has a few more modeling options than **regress**. It still does the basic linear parametric regression. However, the input to the function is the predictor matrix rather than the design matrix. The *predictor matrix* is just the observed values of our predictor variables. In this example, it would be a matrix where the first column contains the latitudes and the second column has the longitude values.

To call **regstats**, use the following syntax

```
stats = regstats(y,X,model)
```

The **model** argument can be one of four strings, and the options are listed here. Note that a constant term is automatically included in all models.

- **'linear'**: First-order terms for each predictor plus a constant
- **'interaction'**: First-order and interaction terms for all predictors with a constant
- **'quadratic'**: First-order, interaction, and squared terms for all predictors, along with a constant term
- **'purequadratic'**: The same as a **'quadratic'** model, but no interaction terms

The default is the **'linear'** model.

There are other **model** options one can use with **regstats**. The model argument can be a matrix that specifies the terms in the polynomial. This matrix is then used with the **x2fx** function to convert the predictor matrix to a design matrix. The **model** matrix has one column for each predictor, and it has one row for each term in the model. The entries in the rows specify the powers for the predictors.

For instance, suppose we had two predictors denoted as X_1 and X_2. Then, we could specify a **'linear'** model using the following **model** matrix:

```
model = [0  0
         1  0
         0  1];
```

Here is an example of a **model** matrix for a dependent relationship with first-order terms, a quadratic term in X_2, and no constant term:

```
model = [1  0
         0  1
         0  2];
```

This gives the user a lot of control over the type of model that can be fit using parametric regression. Note that the **x2fx** function can also be used to create the design matrix for the **regress** function.

TIP

You can call the **regstats** function without any output arguments. In this case, MATLAB provides a GUI where one can select from an extensive list of statistics for the model. These are calculated and returned to the workspace using the specified object names.

The output from the **regstats** function is a structure that contains many useful statistics. A complete list can be found in the documentation. We will illustrate some of them in the next example.

Continuing to use the US temperature data, we will start with the same model as before, which had a first-order term and a constant. However, this time we need a predictor matrix as the input.

```
% We just need the predictor matrix.
Xp = [Lat,Long];
stats = regstats(JanTemp,Xp)
```

Some of the fields in the structure **stats** are shown here. The dots (...) indicate places for other fields that we left out because we will not be discussing them in this primer.

```
stats =

            source: 'regstats'
               ...
              beta: [3x1 double]
               ...
              yhat: [56x1 double]
                 r: [56x1 double]
               ...
               mse: 48.0877
           rsquare: 0.7411
        adjrsquare: 0.7314
               ...
             tstat: [1x1 struct]
             fstat: [1x1 struct]
```

From this, we see that we get most of the same estimates and statistics that we got from **regress**.

The estimated coefficients can be extracted using this expression:

```
% Extract the estimated coefficients.
bhat = stats.beta

bhat =

    98.6452
    -2.1636
     0.1340
```

Similarly, we can get the estimated responses at the observed predictor values using **stats.yhat**. One output that we had before that is not in the **stats** structure is the 95% confidence intervals for the coefficients.

We already discussed the F statistic and associated hypothesis test, and we see a field in **stats** called **fstat**. This is another structure that has the following information about the F-test, as displayed here.

```
>> stats.fstat

ans =

      sse:  2.5486e+03
      dfe:  53
      dfr:  2
      ssr:  7.2973e+03
        f:  75.8753
     pval:  2.7915e-16
```

The value in **stats.fstat.sse** is the sum of squares for the residuals, and **stats.fstat.ssr** is the sum of squares from the regression. Add these together, and we get the total variability of the responses y_i.

There is also a t-test for the coefficients that we can get from **stats**. This provides information to test whether or not the estimated coefficients are significant, which is a way to assess the suitability of the model. Recall that the F-test looks at the model as a whole, while the t-tests focus on individual coefficients. The results of our t-tests are shown here.

```
>> stats.tstat

ans =

     beta:  [3x1 double]
       se:  [3x1 double]
        t:  [3x1 double]
     pval:  [3x1 double]
      dfe:  53
```

The object **stats.tstat.t** has the observed values of our test statistic, and **stats.tstat.pval** has the associated p-values. Here are the p-values.

```
>> stats.tstat.pval

ans =

     0.0000
     0.0000
     0.0386
```

These are all less than 0.05, which means that they are significant at the 5% level and should be kept in the model.

TIP

There is a function called **linhyptest**, which takes values from the output structure **stats** and will perform a linear hypothesis tests. This function can also be used with outputs from other MATLAB model-building functions, such as **nlinfit** and **glmfit**.

6.3.3 The Linear Regression Model Class

The last model-building functions we cover are **LinearModel.fit** and the simpler alternative **fitlm**. The first one became available in MATLAB version R2012a, and **fitlm** was introduced in version R2013b.

We first discuss the options for **LinearModel.fit**. The basic expression to call this function is

```
mdl = LinearModel.fit(X,y,model)
```

where **y** is the vector of observed responses, and **X** is the predictor matrix. One can also use a dataset array **ds** for the input data. In this case, we have

```
mdl = LinearModel.fit(ds,model)
```

Recall from Chapter 1 that a dataset array is a special data object, which is available in the Statistics Toolbox. The rows of a dataset array correspond to the observations. The columns represent the variables, and they usually have variable names attached to them. When building models, we have to specify what column has the responses. The response is the last column of the dataset array by default, but one can change it by specifying the response variable in the function call.

There are several options for **model**. This can take on the same values that we listed for **regstats**:

```
'constant','linear','quadratic','purequadratic'
```

We can specify a general polynomial model with the string **'polyijkm'**, where the highest degree of the first predictor is given by **i**, the highest degree of the second predictor is **j**, and so on. The degree index can take on values 0 to 9. The model will include the highest-order term and all lower-order terms.

An easier and more intuitive way to specify the model can be done using a formula. These formulas have the following general form

```
'Y ~ term1 op1 term2 op2...'
```

Y denotes the name of the response variable. The names of the predictor variables are given by **term1, term2**, and so on. The operations for these terms are specified by **op1**, etc. The operations can be any of the following:

- An operation of **+** means to include the term.
- An operation of **–** means to exclude the term.
- A colon (**:**) operation specifies an interaction or product of two terms.
- The ***** operation will include an interaction and all lower-order terms.
- The **^** operator will include a term with the specified power, and it also includes lower-order terms.

TABLE 6.2

Formula Examples to Specify Models

Formula	Meaning
`'Y ~ -1 + x'`	Straight line fit through the origin. There is no constant term.
`'Y ~ x^2'`	Quadratic polynomial with all terms up to the squared term.
`'Y ~ x1 + x2'`	Multiple regression with first-order terms. It includes the constant term by default.
`'Y ~ x1 + x2 + x1:x2'`	Multiple regression with first-order terms, an interaction term, and a constant.

The models automatically include a constant term when we use either the model name or the formula notation to specify the model. In the first case, we can remove the constant term in the model by setting the **'Intercept'** parameter to **false**. We can exclude the constant in the second case by using a – 1 in the formula. Some examples of formulas are given in Table 6.2.

The **fitlm** function is an alternative to **LinearModel.fit**, and it has similar options and input arguments. The predictor matrix **X** is specified first in **fitlm**. The **fitlm** function will also take a dataset array **ds** as input. Models are specified in the same way as we had with **LinearModel.fit**. The main call to **fitlm** is shown here

$$\text{lm} = \text{fitlm}(X, y, \text{model})$$

The output from **LinearModel.fit** and **fitlm** is an instance of a linear regression model object class, rather than a structure or individual array, as we had with **regstats** and **regress**. Recall that object classes have methods (or special functions) associated with them. We can use the methods for a linear regression object to conduct hypothesis tests, to construct confidence intervals, to create diagnostic plots, and more. We will illustrate some of these methods in the examples that follow.

We continue to use the US temperature data for the examples. We can get the same model as before—a first-order term with a constant—using this code.

```
% Fit the same model as before.
mdl = LinearModel.fit(Xp,JanTemp,'linear');
% Display the model.
disp(mdl)
```

The last expression displays the following table in the command window.

```
Linear regression model:
    y ~ 1 + x1 + x2

Estimated Coefficients:
                   Estimate      SE        tStat       pValue

    (Intercept)     98.645      8.3271     11.846     1.6236e-16
    x1             -2.1636      0.1757    -12.314     3.5692e-17
    x2              0.13396     0.063141    2.1216    0.03856

Number of observations: 56
Error degrees of freedom: 53
Root Mean Squared Error: 6.93
R-squared: 0.741,   Adjusted R-Squared 0.731
F-statistic vs. constant model: 75.9, ...
        p-value = 2.79e-16
```

This is a nice summary that shows the regression model in the formula notation, which allows us to check the model specification. The table also includes estimates of the coefficients and their standard errors, as well as the t-statistics and p-values. The small p-values associated with the estimated coefficients indicate that the terms are significant at the $\alpha = 0.05$ level. At the bottom of the summary, we have the R-squared values and the results of the F-test for the overall model fit. Note that this matches the results we saw before from the **regress** and **regstats** functions.

Now, let's get the same multiple regression model using **fitlm**. We first create a dataset object to show how it is done, and then we fit the model using the dataset object as the data input.

```
% Create a dataset array from the variables.
% Put the response as the last column.
ds = dataset(Lat,Long,JanTemp);
lm = fitlm(ds,'linear');
disp(lm)
```

The summary we have here is a little nicer because we used a dataset object for input. The displayed formula uses the actual variable names instead of generic labels. This is what results from the **disp(lm)**.

```
Linear regression model:
    JanTemp ~ 1 + Lat + Long


Estimated Coefficients:


Estimate          SE           tStat          pValue
               _____       _____        _____        _____


    (Intercept)    98.645         8.3271         11.846      1.6236e-16
    Lat           -2.1636         0.1757        -12.314      3.5692e-17
    Long           0.13396        0.063141        2.1216      0.03856

Number of observations: 56,
Error degrees of freedom: 53
Root Mean Squared Error: 6.93
R-squared: 0.741,   Adjusted R-Squared 0.731
F-statistic vs. constant model: 75.9, ...
    p-value = 2.79e-16
```

An alternative summary can be obtained using the **anova** method. This provides different information from the above. It also helps us examine the quality of the estimated model.

```
tbl = anova(lm)

tbl =
                SumSq      DF     MeanSq       F          pValue
               _____     __    _____    _____     _____


    Lat         7291.7     1      7291.7     151.63      3.5692e-17
    Long        216.46     1      216.46      4.5014      0.03856
    Error       2548.6     53      48.088
```

Finally, we can easily get the 95% confidence intervals for the estimated coefficients, as shown here.

```
bint = coefCI(lm)

bint =

    81.9432     115.3472
    -2.5160      -1.8111
     0.0073       0.2606
```

There are many more methods that one can use with the linear regression model objects, and we summarize the main ones in Table 6.3.

TABLE 6.3

Some Methods for the Linear Regression Model Object Class

Function Name	Functionality
anova	Display an analysis of variance table
coefCI	Create confidence intervals for the coefficients
coefTest	Conduct a linear hypothesis test on coefficients
disp	Display the regression model
feval	Evaluate the regression model for given values
plot	Construct a scatter plot with regression or an added value plot
plotDiagnostics	Create a diagnostic plot based on the model
plotResiduals	Plot the residuals
predict	Produce estimated responses and 95% confidence intervals based on the fitted model

6.3.4 Assessing Model Fit

We have been fitting the same model for US temperature in the examples we have given so far, and we now turn our attention to ways we can assess how well our model fits the data. We touched on this issue somewhat with the hypothesis tests, but we will explore it further in the following examples. We will also demonstrate some ways we can examine the assumptions we made about the distribution of the errors. This is important because the results of the tests are suspect when these assumptions are violated.

Example—Visualizing the Estimated Model

One step to assess a univariate model is to take a look at a scatter plot of the data, along with the curve representing the predicted values from the model. The **plot** function can be used with the linear regression data object, as shown here for our first-order model of temperature and latitude. We are going to fit the same model as we had in Section 6.2, but we are now using the **fitlm** function with the dataset object as the input.

```
% Fit a straight line between temperature and latitude.
% We are using our dataset object from before.
lm1 = fitlm(ds,'JanTemp ~ Lat');

% Plot the model with the data.
plot(lm1)
```

The resulting model is displayed below.

```
Linear regression model:
    JanTemp ~ 1 + Lat
```

```
Estimated Coefficients:
                  Estimate      SE         tStat       pValue
```

	Estimate	SE	tStat	pValue
(Intercept)	108.73	7.0561	15.409	2.034e-21
Lat	-2.1096	0.1794	-11.759	1.624e-16

```
Number of observations: 56
Error degrees of freedom: 54
Root Mean Squared Error: 7.16
R-squared: 0.719,   Adjusted R-Squared 0.714
F-statistic vs. constant model: 138, ...
      p-value = 1.62e-16
```

The results from **plot** are shown in Figure 6.6. The plot includes the data, the estimated model, and the 95% confidence bounds for the predicted values.

The model shown above is a straight line fit between the temperature and the latitude. The R-squared is 0.719, and the adjusted R-squared is 0.714. So far, we showed these values in the summary display of the regression model object, but we can also get them from the linear model object, as shown here.

```
lm1.Rsquared

% This is displayed in the window.
ans =
    Ordinary: 0.7192
    Adjusted: 0.7140
```

These two R-squared values are close together, indicating that our current model is not overfitting the data. This is not surprising because we have only one term in the model.

Example—Assessing Model Appropriateness

Now, we look at the F-test for this model. Recall that this tests the global hypothesis that all of the coefficients β_i, $i = 1, \ldots$ (i.e., excludes the constant term) are zero versus the null hypothesis that at least one of the coefficients

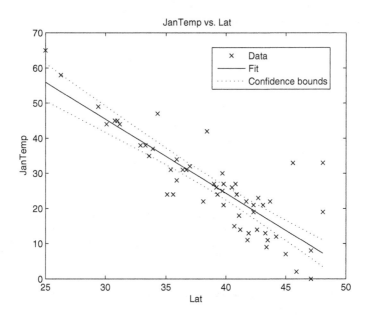

FIGURE 6.6

*This is the plot that is produced when using the **plot** function with a linear regression model object. In the univariate case, it constructs a scatter plot of the data and adds the fitted model, along with 95% confidence bounds.*

is not zero. In other words, the null hypothesis corresponds to a model with just the constant term, and the F-test determines the significance of the overall regression model. The F-statistic and associated p-value are displayed with the model object and when we use the **anova** method. Additionally, we can get the results of the F-test using the following expression. Note that the results are the same as those that are displayed with the model summary.

```
coefTest(lm1)

% This is shown in the command window.
ans =

    1.6240e-16
```

It should not be a surprise that the F-test is significant, given the strong relationship between temperature and latitude (see Figure 6.3). Now, let's see what happens when we construct a model of January temperatures and use a predictor that we know is not related to temperature. To do this we will generate some standard normal random variables, and use these as the predictors for **fitlm**.

```
% This time we use the predictor matrix and vector
% of responses as inputs to fitlm.
% Generate some normal RVs as a predictor variable.
xn = randn(size(JanTemp));

% Fit a first-order model.
lm2 = fitlm(xn, JanTemp, 'linear');
disp(lm2)
```

The resulting model is summarized here.

```
Linear regression model:
    y ~ 1 + x1

Estimated Coefficients:
                  Estimate      SE      tStat       pValue

    (Intercept)    26.579     1.812    14.668     1.7564e-20
    x1             0.55165    1.6642   0.33149       0.74156

Number of observations: 56
Error degrees of freedom: 54
Root Mean Squared Error: 13.5
R-squared: 0.00203,   Adjusted R-Squared -0.0165
F-statistic vs. constant model: 0.11, ...
      p-value = 0.742
```

Note the large p-value of 0.742 for the F-test and the very small R-squared. The F-test correctly indicates that we should accept the null hypothesis of a constant model or relationship between the response and predictor. The small R-squared indicates a similar lack of relationship.

We continue to explore this idea further by adding the same predictor to the model of **JanTemp** and **Lat**.

```
% Add the predictor that is just random
% noise and fit a multiple regression with first-order
% terms for Latitude and the noise predictor.
% Include latitude and noise in the predictor matrix.
Xn = [Lat, xn];
lm3 = fitlm(Xn,JanTemp,'linear');
disp(lm3)
```

Here is a summary of the model and the associated hypothesis tests.

```
Linear regression model:
    y ~ 1 + x1 + x2

Estimated Coefficients:
```

	Estimate	SE	tStat	pValue
(Intercept)	110.08	6.9568	15.823	9.9655e-22
x1	-2.1399	0.17664	-12.114	6.7894e-17
x2	1.5582	0.86922	1.7926	0.07874

```
Number of observations: 56
Error degrees of freedom: 53
Root Mean Squared Error: 7.01
R-squared: 0.735,   Adjusted R-Squared 0.725
F-statistic vs. constant model: 73.6, ...
      p-value = 5.09e-16
```

The p-value of our F-statistic is very small, indicating that at least one of our coefficients (excluding the constant) is not zero.

Now, let's look at the t-test for the coefficients of the terms in the model. The p-value corresponding to the latitude term is very small, which means that the predictor is significant and should be kept in the model. However, the p-value for the noise predictor (**x2**) is greater than 0.05, and we do not have significant evidence at the 5% level that $\beta_2 \neq 0$. This is an indication that we can remove that predictor (or term) from our model.

We can also look at 95% confidence intervals for the estimated coefficients, as we show below.

```
coefCI(lm3)

% The following coefficients are displayed.
ans =

   96.1262   124.0333
   -2.4941    -1.7856
   -0.1852     3.3016
```

From this, we see that the confidence interval for β_2 is $[-0.1852, 3.3016]$, which contains zero. This is an alternative way to perform the same test on the significance of the coefficient. Alternatively, the confidence interval for β_1 does not contain zero, which means that this term is significant.

Finally, we could look at the R-squared values for the two models—**lm1** and **lm3**. The first model has R-squared values of 0.719 and 0.714 (adjusted), while the model with the noise predictor has values of 0.735 and 0.725, as shown here.

```
lm3.Rsquared

% The R-squared values are shown in the window.
ans =
```

```
Ordinary: 0.7352
Adjusted: 0.7252
```

The unadjusted R-squared values for the two models increased, as expected. However, we see that the adjusted R-squared for **lm3** did not have the same increase as the ordinary R-squared. This is another indication that the additional term (**xn**) in the model might not belong there.

We now turn our attention to the model we fit previously using the two predictors—latitude and longitude. As you might recall, we suspected that a straight line fit between temperature and longitude might not be a good one; see Figure 6.5. In this example, we are going to fit a first-order model (**mdl1**) using additive terms for **Lat** and **Long**. The second model **mdl2** will have the same first-order terms with an additional quadratic term for **Long**. Finally, the third model (**mdl3**) has the same terms as **mdl2** with a cubic term on **Long**. This produces a sequence of nested models, and we can use the R-squared values to compare them. Here is the MATLAB code to create these models.

```
% Adjust the model by adding terms for longitude.
% Repeating the linear model first...
mdl1 = fitlm(ds,'JanTemp ~ Lat + Long');

% Next, add a quadratic term in longitude.
mdl2 = fitlm(ds,'JanTemp ~ Lat + Long^2');

% Now, add a cubic term in longitude.
mdl3 = fitlm(ds,'JanTemp ~ Lat + Long^3');

% Get the R-squared values.
R1 = mdl1.Rsquared
R2 = mdl2.Rsquared
R3 = mdl3.Rsquared
```

Here are the R-squared values for the models.

```
R1 =
     Ordinary: 0.7411
     Adjusted: 0.7314

R2 =
     Ordinary: 0.8724
     Adjusted: 0.8651

R3 =
     Ordinary: 0.9461
     Adjusted: 0.9419
```

As expected, the ordinary R-squared values increased as we added terms to the model, but there is a gap between the ordinary and adjusted R-squared values. This indicates that the additional terms in **md11** and **md12** might not be highly explanatory. However, the difference between the R-squared values for the last model is very small, and this is an indication that the terms belong in the model. Furthermore, our model (**md13**) is now explaining 94% of the variability in temperature.

Example—Assessing the Assumptions

We conclude this section with a brief discussion and some examples on how to check the assumptions of our model. We will use the last model to show how to determine the distribution of our errors and assess the constant variance assumption. We will do this using visualization.

We first look at how the errors from our model are distributed by looking at the residuals. The residuals are an estimate of the errors and are given by

$$\hat{\varepsilon}_i = y_i - \hat{y}_i.$$

We can construct a probability plot using the **plotResiduals** function, as shown here.

```
% Construct a probability plot of the residuals.
plotResiduals(md13,'probability')
```

This compares the residuals to a standard normal distribution. The plot for our model is shown in Figure 6.7 (top), and we see that, while the points fall on the line for the most part, there is some evidence of potential outliers or non-normality in the tails.

We also assumed that the errors have constant variance. We can verify this by looking at a scatter plot of the residuals versus the fitted values. If the constant variance assumption is correct, then we would expect the scatter plot to show a horizontal band of points, with no patterns or trends showing. We can get this plot using the **'fitted'** option, as shown here. The plot is shown in Figure 6.7 (bottom).

```
% Plot of the residuals against the fitted values.
plotResiduals(md13,'fitted')
```

TIP

The default for the **plotResiduals** function is to plot a histogram of the residuals. This can be used to determine how the residuals (or the estimated errors) are distributed and also to locate possible outliers.

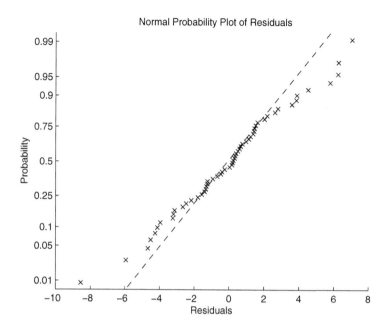

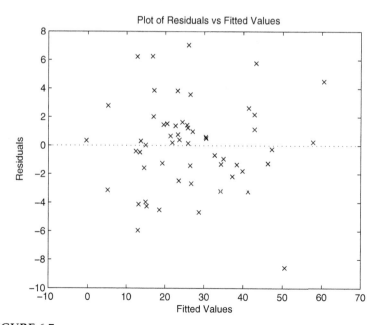

FIGURE 6.7

The top figure is a normal probability plot of the residuals, and it is used to check the distribution of the errors. The lower one shows the residuals against the fitted values. The horizontal band of points indicates our assumption of constant variance is reasonable.

6.4 Basic Fitting GUI

There is a graphical user interface (GUI) capability for basic model fitting that is provided with the base MATLAB package, and it can be accessed using the Figure window TOOLS menu. It is called the Basic Fitting GUI.

We return to the data used in our initial example in this chapter to show how to use the Basic Fitting GUI. Recall that we generated some noisy data according to the following true underlying quadratic relationship

$$y = -2x^2 + 2x + 3.$$

See Section 6.2.1 for the MATLAB code that created the data.

The tool works with univariate data only, when we have one predictor and one response variable. The tool can be accessed once a Figure window is opened. Therefore, the first step is to construct a scatter plot.

```
% First construct the scatter plot.
% The Basic Fitting GUI can be accessed in the
% Tools menu of the Figure window.
plot(x,y,'.')
```

To open the GUI, click on the TOOLS menu, and then select the BASIC FITTING option at the bottom. The scatter plot of our data and the TOOLS menu is shown in Figure 6.8.

This produces the GUI shown in Figure 6.9. The large area in the middle has many model-fitting choices that range from interpolants to a polynomial of degree ten. Below this are some choices for displaying the residuals. One can display them as a bar plot, a scatter plot, or a line plot. There are also options for showing the residual plots in a subplot along with the data or in a separate Figure window. Finally, the button in the lower right corner that has an arrow on it will expand the GUI to include a text window that will display the fitted model. See Figure 6.10 for an illustration.

If you select one of the interpolants for the model, then the resulting curve will fit the data exactly. In other words, the curve passes through each data point. This is also what happens when one plots the (x_i, y_i) values in MATLAB as a line. The points are plotted and then each one is connected with a straight line. In this case, the curve or model exactly fits the points and the residuals are zero.

We show an example of this in Figure 6.10, where we chose the spline interpolant. We expanded the GUI to see the model and information about the residuals. In this case, the model is a spline interpolant, and the norm of the residuals is zero, as expected. The model is also superimposed on the plot of the data.

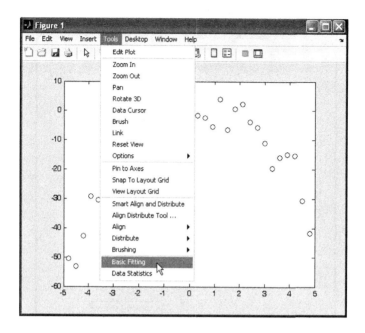

FIGURE 6.8
We first create a scatter plot of our data. We then click on the **Tools** *menu and select* **Basic Fitting***.*

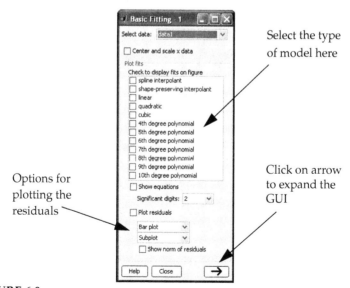

FIGURE 6.9
This is what appears when the Basic Fitting GUI is started. There are many options for the type of model to fit, ranging from interpolants to polynomials of degree 10.

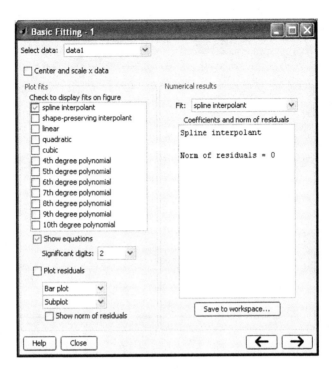

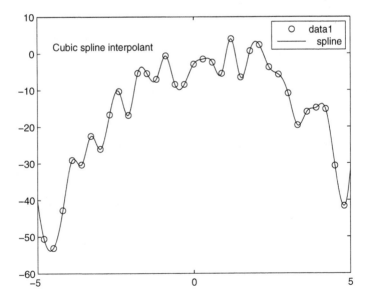

FIGURE 6.10
The top figure is a screenshot of the Basic Fitting GUI showing the expanded area that provides information about the model. We chose a spline interpolant, which is displayed on the plot. Note that it fits the data exactly. Thus, the norm of the residuals is zero.

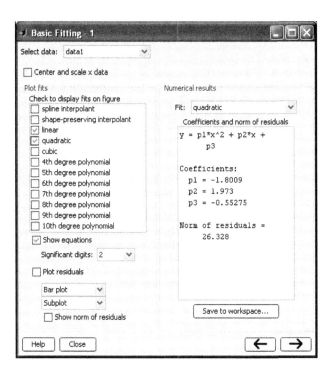

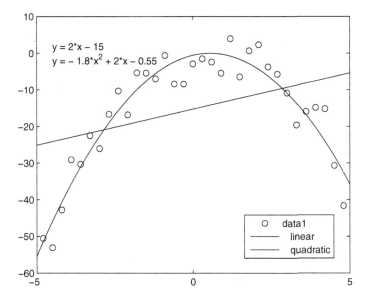

FIGURE 6.11

The estimated quadratic model is shown in the top figure, along with the norm of the residuals. The estimated linear and quadratic models are shown on the plot. The quadratic model fits the data better than the straight line.

We now demonstrate what happens when we select two models—linear and quadratic. The linear option in the GUI corresponds to a model with a first-order term and a constant. The quadratic is the second choice we clicked on, so that is the one shown in the expanded GUI window in Figure 6.11. Both estimated models are shown as curves on the scatter plot. We selected SHOW EQUATIONS on the GUI, and these are also displayed on the scatter plot.

There is a button on the right side of the GUI that says SAVE TO WORKSPACE. Clicking on this button brings up an additional GUI window that will save the fitted model as a structure, the residuals, and their norm.

> *TIP*
>
> Clicking on the additional right-arrow button at the bottom of the Basic Fitting GUI (see Figures 6.10 and 6.11) will expand the GUI further. This provides an area with options to predict and to plot values of the response for given values of the predictor.

6.5 Summary and Further Reading

We could only give a brief introduction to regression analysis in this chapter. There are many more options available that have been developed in the literature and are also implemented in MATLAB.

The first part of this chapter provided some background information on regression analysis and model building. This included definitions for linear parametric models, an introduction to the ordinary least squares approach to estimate the models, and a description of statistics and hypothesis tests that we can use to assess our estimated models. We then described the functions for regression analysis that are available in the base MATLAB software. These are summarized in Table 6.4.

The next section focused on functions for linear parametric regression in the Statistics Toolbox. There are several more approaches to model-building listed in Table 6.5. These include robust fitting methods that are resistant to outliers, stepwise regression for model development, and generalized linear models for situations when one has restrictions on the responses (e.g., binary responses).

We defined nonparametric regression in the beginning of this chapter, but we did not provide further information on approaches for estimating models of this type. Recall that in this case, we do not assume a specific parametric model and are not estimating coefficients. Table 6.6 lists some of the functions for nonparametric regression in the Statistics Toolbox. MATLAB has an additional toolbox for nonparametric using smoothing methods, such as loess and smoothing splines. These are in the Curve Fitting Toolbox.

TABLE 6.4

Model-Building Functions in Base MATLAB®

Function Name	Description
`\ , mldivide`	Left-divide operator, solve system of equations given by $\mathbf{Ax} = \mathbf{b}$
`/ , mrdivide`	Right-divide operator, solve system of equations given by $\mathbf{xA} = \mathbf{b}$
`linsolve`	Solve system of equations $\mathbf{Ax} = \mathbf{b}$
`lscov`	Solve system of equations using least squares or weighted least squares
`polyfit`	Fit a polynomial of degree p to one predictor

TABLE 6.5

Parametric Model-Building Functions in the Statistics Toolbox

Function Name	Description
`fitlm`	Estimate coefficients in a linear model
`glmfit`	Fit a generalized linear model
`lasso`	Use lasso regression approach to fit a linear model
`LinearModel.fit`	Estimate coefficients in a linear model
`mnrfit`	Fit a multinomial regression model
`mvregress`	Multivariate regression—multiple responses
`plsregress`	Use partial least squares to estimate model
`regress, regstats`	Fit linear model using parametric regression
`ridge`	Use ridge regression to estimate linear model
`robustfit`	Estimate a model that will deal with outliers
`stepwiselm`	Fit parametric models using stepwise regression

TABLE 6.6

Nonparametric Model-Building Functions in the Statistics Toolbox

Function Name	Description
`fitnlm`	Fit a model with nonlinear coefficients
`fitensemble`	Estimate a model using ensemble methods, such as boosting and bagging
`nlmefit`	Fit a nonlinear mixed-effects model
`RegressionTree.fit` `fitrtree`	Estimate a model using regression trees

Regression analysis using basic least squares is included in most books on statistics, engineering, applied mathematics, and linear algebra. The reader is encouraged to consult any of these for more information on this fundamental data analytic tool. However, for more details, we recommend *Regression Graphics* by Cook [1998] and *Applied Regression Analysis* by Draper and Smith [1981].

The online documentation for base MATLAB has a section on estimating models. It can be found under the *Mathematics* chapter and the *Linear Algebra* section. The main chapter for model-building in the Statistics Toolbox is called *Regression and ANOVA*. Finally, the interested reader can refer to the section on *Classification Trees and Regression Trees* under the *Machine Learning* chapter for details on the tree approach for regression analysis. There is also a section on *Nonlinear Regression* in that chapter.

Chapter 7

Multivariate Analysis

Data do not give up their secrets easily. They must be tortured to confess.
Jeff Hooper, Bell Labs

Today's data sets tend to be massive in terms of the number of observations n and the number of variables or characteristics p that we measure. Therefore, we often need to reduce the number of variables through dimensionality reduction or feature transformation before we can analyze our data. This will be the topic of the first two sections in this chapter, where we cover principal component analysis and multidimensional scaling. In the last section, we will present ways to visualize high-dimensional data before any transformation is done. These include scatter plot matrices, graphing in parallel coordinates, and Andrews' curves.

7.1 Principal Component Analysis

Principal component analysis or PCA is a method for transforming data based on eigenvalue analysis. It is one of the key techniques from linear algebra. In statistical analysis, there can be two uses for PCA. The first one is to transform the data to a new set of variables that are a linear combination of the original ones. The result is the same number of variables—they are just in a different space. The second goal is to reduce the dimensionality from p to d, where $d < p$. We would like to do this in a way that accounts for some percentage of the variation in the original data.

As we will see in the examples, functions for PCA in MATLAB® use either the original data matrix **X**, the covariance matrix, or the correlation matrix. The usual way of explaining PCA is based on the data covariance matrix, which we will denote as **S**. The ij-th element of **S** is the covariance between the i-th and j-th variables, and the diagonal elements of the matrix hold the variances.

PCA is based on eigenanalysis of matrix **S**. This uses a general result from linear algebra showing that any square, symmetric, nonsingular matrix can be converted into a diagonal matrix using the results from eigenanalysis. A square matrix has the same number of rows and columns, and a symmetric matrix is one that satisfies

$$\mathbf{S} = \mathbf{S}^T.$$

The superscript T indicates the transpose of a matrix, where the rows and columns are switched. A nonsingular matrix is one that has an inverse, or alternatively, one that has a nonzero matrix determinant. Note that the covariance matrix from the observed data is both square and symmetric.

The eigenvalues λ_j are found by solving this matrix equation

$$|\mathbf{S} - \lambda\mathbf{I}| = \mathbf{0},$$

where **I** is the $p \times p$ identity matrix. The vertical lines denote the matrix determinant. Each eigenvector $\mathbf{a}_j$, $j = 1, ..., p$, is obtained by solving

$$(\mathbf{S} - \lambda_j\mathbf{I})\mathbf{a}_j = \mathbf{0},$$

Thus, $\mathbf{a}_j$ is the eigenvector associated with the j-th eigenvalue λ_j.

There are constraints imposed on the eigenvectors. They are orthogonal to each other, and each of them has a magnitude of one. Therefore, they provide an orthonormal basis for the space given by the data, and we can transform the data into this space. The transformed data have some nice properties we will demonstrate in the next section.

The new variables are called ***principal components*** (PCs). We will denote the transformed variables as z_j, and they are found using

$$z_j = \mathbf{a}_j^T\mathbf{x},$$

with **x** representing the original p variables. From this, we see that the PCs are a linear combination of our initially measured characteristics represented by **x**. The elements of the eigenvectors provide the weights or the contribution of the original variables in the new principal coordinate space.

We can transform the data to the principal component space using matrix multiplication, as shown here

$$\mathbf{Z} = \mathbf{X}\mathbf{A},$$

where **X** is our $n \times p$ data matrix, and **A** is a $p \times p$ matrix with each column containing a principal component or eigenvector. The transformed data in the matrix **Z** are called ***principal component scores***.

The convention is to order the columns of $\mathbf{A}$ in descending order of the eigenvalues, $\lambda_1 \geq \ldots \geq \lambda_p$. In other words, the first column of $\mathbf{A}$ is the eigenvector that was obtained using the largest eigenvalue, the second from the next largest eigenvalue, and so on. We can reduce the dimensionality of the data by using only the eigenvectors that have the largest eigenvalues, as shown here

$$\mathbf{Z}_d = \mathbf{X}\mathbf{A}_d,$$

where the d columns of $\mathbf{A}_d$ contain the first d eigenvectors. The matrix $\mathbf{A}_d$ has dimensions $p \times d$, and we see that our transformed data matrix $\mathbf{Z}_d$ now has only d columns.

The principal component scores (the transformed data) are uncorrelated, and the variance of each principal component is equal to its corresponding eigenvalue. The other nice property of PCA is that the sum of the eigenvalues equals the sum of the variances in the original data. Thus, if we use all of the eigenvectors in the transformation, then we are representing all of the variation in the original data. If we use the d largest eigenvalues, then we are accounting for some smaller percentage of the original variation.

We can think of it this way. Each principal component defines an axis or coordinate in a new space. If we project the data to the coordinate axis that corresponds to the largest eigenvalue, then we are accounting for the maximum amount of variance possible using one dimension. If we include a second principal component in the transformation, then we are accounting for the maximum amount of the remaining variation, and so on.

We can use this concept to help us decide how many principal components to keep or alternatively what value to use for d. The most common way to determine the number of PCs is to select the d components that contribute a given percentage of the total variation in the data. This can be determined as

$$t_d = 100 \times \sum_{j=1}^{d} \lambda_j \div \sum_{j=1}^{p} \lambda_j.$$

There is a graphical way to help choose the number of PCs to retain. It is called the *scree plot* [Cattell, 1966]. This plots the ordered eigenvalues against their index. We look for an elbow in the curve to estimate the number of dimensions one should keep. The curve typically becomes flat after the elbow, implying that the remaining variance as given by the eigenvalues is less.

In the next section, we describe the options for eigenanalysis that are in the base MATLAB software, and we provide some examples illustrating the concepts we just described. This is followed by examples of functions for PCA that are provided in the Statistics Toolbox.

TIP

If the eigenvalues are extremely large, then we can plot the log of the eigenvalues against the index in the scree plot.

7.1.1 Functions for PCA in Base MATLAB®

There is a function for general eigenanalysis in the base MATLAB software. It is called **eig**, and it takes a square matrix as an input argument. The basic syntax to get the eigenvectors and eigenvalues is

$$[A,D] = eig(B)$$

The input matrix **B** is a square matrix. The output matrix **A** contains the eigenvectors as columns, and the matrix **D** is a diagonal matrix with entries given by the eigenvalues. It is important to note that the eigenvalues we get from **eig** are typically given in *ascending* order, which is the opposite of the convention mentioned earlier.

Our first example illustrating the **eig** function uses data that we generate according to a multivariate normal centered at $(2, -2)$. It has a covariance matrix given by

$$\Sigma = \begin{bmatrix} 0.9 & 0.4 \\ 0.4 & 0.3 \end{bmatrix}.$$

The MATLAB code to generate the variables is shown below.

```
% Generate some multivariate Normal data
% Set up a vector of means.
mu = [2 -2];

% Specify the covariance matrix.
Sigma = [.9 .4; .4 .3];
X = mvnrnd(mu, Sigma, 200);
plot(X(:,1),X(:,2),'.');
xlabel('X_1')
ylabel('X_2')
title('Correlated Multivariate Normal Data')
```

The scatter plot is shown in Figure 7.1.

We are ready to perform the eigenanalysis of the covariance matrix based on the data. We first get the covariance matrix, and then we verify that the matrix satisfies the requirements for the analysis.

```
% First get the covariance matrix.
% This will be the input to eig.
covm = cov(X);

% Get the number of rows and columns
size(covm)

ans =

    2    2
```

The **size** function tells us that the matrix has two rows and columns, so it is a square matrix.

Next, we see if it is a symmetric matrix. It has to be symmetric by definition, but this shows how to verify that.

```
% Demonstrate that this is symmetric. If it is,
% then the transpose is equal to itself.
isequal(covm,covm')

% A result of 1 indicates TRUE.
ans =

    1
```

The result from **isequal** is 1, indicating that the transpose of **covm** is equal to itself, and thus, is symmetric.

A matrix is nonsingular if the inverse exists, which also means that the determinant is nonzero. As we will see, we can also find out whether a matrix is nonsingular by looking at the eigenvalues. Let's find out if the inverse exists for our covariance matrix.

```
% It is nonsingular because the inverse exits.
covmI = inv(covm)

covmI =

    2.8607   -3.8426
   -3.8426    8.1656

% What is the determinant?
det_cov = det(covm)

det_cov =

    0.1164
```

Thus, we see that our covariance matrix is also nonsingular.

We can now call the **eig** function to get the eigenvectors and eigenvalues.

```
% Next get the eigenvalues and eigenvectors.
[A,D] = eig(covm);
```

The eigenvalues are shown here as the diagonal elements of **D**, and we see that they are in ascending order.

```
% Display the elements of D.
display(D)

D =

    0.0982        0
         0    1.1849
```

We can transform the data from the original space to the space spanned by the principal components, as given below. We also create a scatter plot of the data in the new space, which is shown in the bottom of Figure 7.1.

```
% Now project onto the new space and plot.
% Just data transformation at this point.
Z = X*A;
plot(Z(:,1),Z(:,2),'.')
xlabel('PC 1')
ylabel('PC 2')
title('Data Transformed to PC Axes')
```

We can tell from the shape of the data cloud that they are correlated in the original variable space. The data are now uncorrelated when transformed to the new variables based on PCA. We can check that by finding the covariance of the transformed data.

```
cov(Z)

% This is displayed in the command window.
ans =

    0.0982   -0.0000
   -0.0000    1.1849
```

The off-diagonal elements are zero, as expected.

We can also verify some of the properties we mentioned regarding PCA. The variances in our PCA space are given by the diagonal elements of the covariance matrix shown above—0.0982 and 1.1849. These are the same as the eigenvalues, as given in the diagonal matrix **D** above.

We also stated that the sum of the eigenvalues is equal to the sum of the variances. We can check that, also.

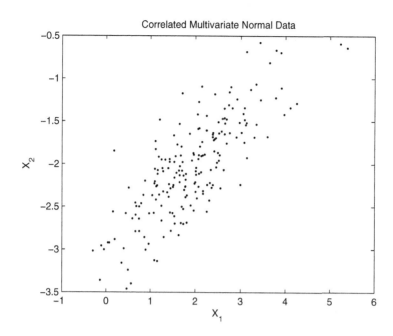

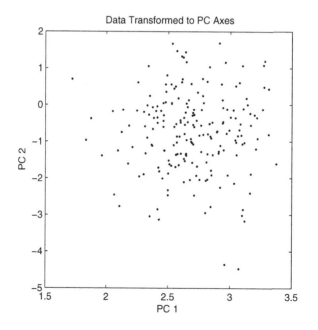

FIGURE 7.1
The first plot shows the data in the original space. These are correlated multivariate normal random variables. We used PCA to transform them to a space where they are now uncorrelated.

```
% Get the sum of the original variances
sum(diag(covm))

ans =
      1.2831

% Get the sum of the eigenvalues
sum(diag(D))

ans =
      1.2831
```

These are the same total variance.

TIP

An alternative test for a matrix to be nonsingular is based on the eigenvalues of a matrix. The product of the eigenvalues is equal to the determinant. Therefore, a matrix is nonsingular if all eigenvalues are nonzero. Execute this command: **prod(diag(D))**. Compare it to the value of **det(covm)** for the previous example.

7.1.2 Functions for PCA in the Statistics Toolbox

The **eig** function in base MATLAB is appropriate for eigenanalysis of any valid input matrix, not just the covariance matrix. There are two functions for PCA in the Statistics Toolbox—**pca** and **pcacov**. These perform an analysis from a statistical perspective.

The **pca** function takes the raw data matrix **X** as an input. The function call to get the eigenvectors, transformed observations, and the eigenvalues is

$$[\text{coeff},\text{score},\text{vals}] = \text{pca(X)}$$

The **coeff** output contains the principal components or eigenvectors, and they are ordered in *descending order* of the eigenvalues—unlike the function **eig**. The transformed observations are in the matrix **score**, and the eigenvalues are in the **vals** output object. These are also called the *principal component variances*.

The default in **pca** is to center the data before it finds the eigenvectors and eigenvalues. This means that the transformed observations in **score** will be centered at the origin. However, there is a name-value pair that one can specify as an input to the function **pca** to change this step.

MATLAB returns the principal component scores for all p eigenvectors, so no dimensionality reduction is achieved. However, we can always reduce the number of dimensions by extracting the first d columns of the **score** matrix to get our d-dimensional data.

The **pcacov** function takes the covariance matrix of the data as an input instead of the raw data. It is not as flexible as the **pca** function in terms of the options that the user can specify, and it has slightly different outputs than **pca**. For example, it does not return the transformed data. Finally, **pcacov** does not center the data, because the input is the covariance matrix, not the raw data. So, the principal component scores that one gets are not centered at the origin, as is the case with the **pca** function.

The syntax for **pcacov** is shown here

```
[coeff,vals,var_exp] = pcacov(covm)
```

The first two outputs are the same as we have with **pca**. The third output is a vector showing the percentage of the total variance explained. We can get the principal component scores by using the same calculation we did with **eig**. This is illustrated in the next examples.

Example of the **pca** Function

We are going to use Fisher's iris data for these next examples. We first load the **iris.mat** file. This imports three variables into the workspace: **setosa**, **versicolor**, and **virginica**. We put them into a data matrix **X**.

```
% Load the iris data and
% put into one matrix.
load iris
X = [setosa; versicolor; virginica];
```

We apply the **pca** function that takes the data matrix as input, and we get the principal component scores as one of the outputs.

```
% Use the PCA function that inputs raw data.
% The outputs match previous eigenanalysis notation.
[A_pca,Z_pca,vl_pca] = pca(X);
```

We plot the observations in coordinates given by the first two principal components, as shown in Figure 7.2. The transformed data are centered at the origin because we used the **pca** function with default options. We could keep the data represented only by these first two principal components (as in Figure 7.2) in any subsequent analysis, which means that we have reduced the dimensionality of the data to $d = 2$. We will look at how much of the variance is explained by these two components in the next example.

Example of the **pcacov** Function

We now use the **pcacov** function in the Statistics Toolbox to perform a similar analysis. This takes the covariance matrix as an input, so we calculate that first.

FIGURE 7.2
We constructed a scatter plot of Fisher's iris data that have been transformed to the orthogonal system spanned by the principal components. Note that the data are centered at the origin in the plot, because we used the **pca** *function with the default options.*

```
% Use the PCA function that takes the
% covariance matrix as input
cov_iris = cov(X);
[A_pcacv,vl_pcacv,var_exp] = pcacov(cov_iris);
```

The principal component scores are not returned from **pcacov**, but we can transform the observations using the matrix of eigenvectors **A_pcacv**. The second output above (**vl_pcacv**) contains the eigenvalues. Note that the columns of the eigenvector matrix are given in decreasing order of the eigenvalues.

In calling the **pcacov** function, we asked for the vector that contains the percentage of the variance explained by the principal components. These are in the vector **var_exp** in our example. This is given below:

```
var_exp =

    92.4619
     5.3066
     1.7103
     0.5212
```

This indicates that the first principal component accounts for 92.46% of the variance in that space, and the second one accounts for an additional 5.31%. We can create a scree plot using the following steps.

```
% Create a scree plot.
plot(1:4,vl_pcacv,'-o')
axis([0.75 4 0 4.5])
xlabel('Index')
ylabel('lambda _j')
title('Scree Plot for Fisher''s Iris Data')
```

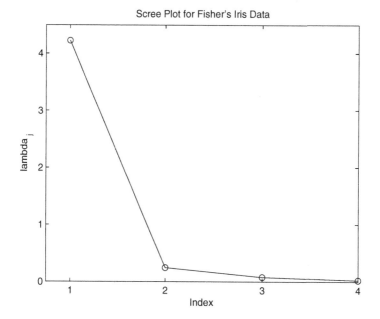

FIGURE 7.3

This shows the scree plot for the eigenanalysis of Fisher's iris data. There is an elbow in the curve at an index of two, which is an indication that we could adequately represent our data in the two dimensions given by the first two principal components.

The plot is shown in Figure 7.3, and we see that there is an elbow in the curve at index (or dimension) two. Therefore, it seems reasonable to represent our data using the first two principal components. According to the entries in **var_exp**, this explains 97.77% of the variance.

```
% The scree plot indicates that we should keep two
% dimensions. This explains the following
```

```
% percentage of the variance.
sum(var_exp(1:2))

ans =

    97.7685
```

TIP

You can use the correlation matrix with **pcacov**, but some results have to be interpreted differently.

7.1.3 Biplots

We conclude this section by discussing a special plot that was developed to convey the results of dimensionality reduction techniques like PCA. This is called the *biplot*. It was developed by Gabriel [1971] as a graphical display of matrices with rank two, and he showed how it could be applied to PCA.

The biplot is essentially a two-dimensional scatter plot of the data in the space given by the first two principal components. The original variables are shown as vectors. This provides two views of the data—the transformed observations as points and the variables as vectors. We obtain the biplot for the first two principal components of Fisher's iris data with the following code. See Figure 7.4 for the resulting biplot.

```
% Get a biplot for the iris data.
varlab = {'S-Length','S-Width',...
    'P-Length','P-Width'};
Z_pcacv = X*A_pcacv;
figure,biplot(A_pcacv(:,1:2),...
    'scores',Z_pcacv(:,1:2),...
    'VarLabels',varlab)
ax = axis; axis([-.2 1 ax(3:4)])
box on
```

The direction and magnitude of the variable vectors show how each one contributes to the principal components displayed in the plot.

7.2 Multidimensional Scaling—MDS

There are several *multidimensional scaling* techniques that can be used to analyze data that are in the form of similarities or distances. The goal of

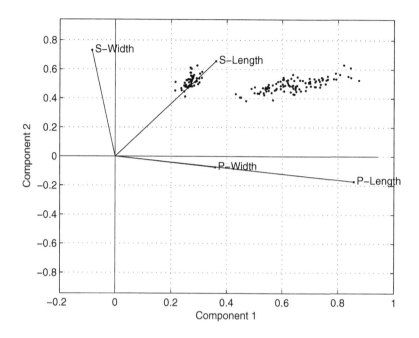

FIGURE 7.4

Here is the biplot in the first two principal components for Fisher's iris data. Comparing this to Figure 7.2, we see the scatter plot of transformed observations (i.e., the principal component scores). However, the scores are scaled, and only their relative positions can be seen in this plot. The four original variables are shown as vectors, where the length of the line indicates the value of the coefficient. The vector is aligned in the direction most strongly related to that variable.

multidimensional scaling (MDS) is to find coordinates or locations of the data in a low-dimensional space that preserve these proximity relationships. Therefore, data points that are close together in the original space defined by the p variables will also be close together in the new configuration.

In some disciplines and applications, we might not have raw data like we have been looking at throughout this primer. Perhaps we only have data about how similar or dissimilar objects are to one another. This is sometimes the case in applications from marketing, sociology, biology, and psychology.

MDS was originally developed in psychometrics to analyze judgments people have about the similarity between objects [Torgerson, 1952], where that was the only data one had to analyze. However, the method can be used with the type of data we have encountered so far and is widely available in statistical software, including MATLAB.

There are two categories of multidimensional scaling (MDS)—metric and nonmetric. The difference between them comes from the assumptions of how the proximities are transformed into the configuration of points in the lower-dimensional space. In what follows, we first define proximities and distances, and how we can get these in MATLAB. This is important for MDS, as well as data clustering, which is described in Chapter 8. After this, we provide information and examples on the three options for MDS in the Statistics Toolbox.

7.2.1 Measuring Distance

In general, the input to MDS is a set of proximities between objects or data points. It could be that this is all the information we have. Or, we could calculate them from the raw data, as we show in an example. A *proximity* indicates how close objects or observations are in space, time, taste, meaning, character, usefulness, etc. How this is defined can impact our analysis and results.

We can measure proximity using similarity or dissimilarity. *Similarity* indicates how alike objects are to one another. A small value of similarity means they are not alike, and a large value means that they are. Sometimes, similarity measures are scaled between zero (not similar) and one (the same).

Dissimilarity measures are the opposite. Small values indicate that they are alike or close together. We will denote dissimilarity between the i-th and j-th objects by δ_{ij}. We note that the dissimilarity of an object with itself is zero ($\delta_{ii} = 0$), and that similarities and dissimilarities are both nonnegative and symmetric. Because one can always convert from one to another [Martinez, et al., 2010], we will use the general term *proximity* to encompass both of these measures.

Another way to measure closeness is by using a *distance measure*. We are going to denote the distance between the i-th and j-th objects as d_{ij}. This should not be confused with the dimensionality d of the transformed data. A proximity measure must meet the following to qualify as a metric distance:

- The distance between an observation and itself is zero.
- Distance is nonnegative.
- It is symmetric: $d_{ij} = d_{ji}$.
- It satisfies the triangle inequality: $d_{ij} \leq d_{ik} + d_{kj}$.

The Euclidean distance is probably the one used most often. It is defined, as follows:

$$d_{ij} = \sqrt{\sum_{k=1}^{p} (x_{ik} - x_{jk})^2} \, ,$$

where x_{ik} is the k-th coordinate of the i-th data point.

There are many options for measuring proximity that can be used with different types of data, such as unstructured text, sets of general objects, and categorical data. Many of them are not metric distances because they do not satisfy the above conditions. See Hartigan [1967] and Cormack [1971] for a taxonomy of proximity functions. We will focus on distances in this book, because the Statistics Toolbox has a function that calculates the interpoint distances between the points in a data set.

We can represent the interpoint distances between all data points in an $n \times n$ matrix. We will denote this by **D** and will call it the *interpoint distance matrix*. The ij-th element of **D** represents the distance between the i-th and j-th observations. The matrix **D** is symmetric due to the symmetry property of distances.

The MATLAB function **pdist** will return the interpoint distances between all n observations. There is some redundant information in the matrix **D** because of the symmetry property. So, **pdist** returns the distances in a vector to save computations and memory. There is a function called **squareform** that will convert them to the full matrix **D**, if necessary. The basic syntax for the function is

$$\textbf{Dv = pdist(X, 'distance')}$$

TABLE 7.1

Distance Options in MATLAB®

Function	Description
`'euclidean'`	Euclidean distance (default)
`'seuclidean'`	Standardized Euclidean distance
`'cityblock'`	City block distance
`'minkowski`	Minkowski distance
`'chebychev'`	Maximum coordinate distance
`'mahalanobis'`	Weighted by the covariance first
`'cosine'`	One minus the cosine similarity
`'correlation'`	One minus the linear correlation
`'spearman'`	One minus the rank correlation
`'hamming'`	Percentage of different coordinates
`'jaccard`	One minus the percentage of different nonzero values
functionname	User-defined distance function

The input **'distance'** is a string that specifies what distance to use. MATLAB provides twelve possibilities (see Table 7.1), including a custom distance. The documentation for **pdist** has the formulas and definitions for the distances. Typing **help pdist** at the command line will provide a list of the options.

The **cosine, correlation, spearman,** and **jaccard** options in **pdist** return measures that have been converted from similarities to distances. We now illustrate the use of **pdist** using the location of the 56 US cities in the **UStemps** data.

Example—Getting Distances

We first load the data into the workspace. This will give us the locations of the cities in longitude and latitude. The next step is to put the locations into a matrix that we can use as input to **pdist**. We also create a scatter plot of the cities shown in Figure 7.5. We add some city labels in the **City** object to provide context.

```
load ustemps
% Put the locations into a data matrix.
X = [-Long(:),Lat(:)];

% Plot the latitude and longitude as a scatter plot.
plot(-Long,Lat,'*')
axis([-130 -65 23 50])
xlabel('Longitude'), ylabel('Latitude')
title('Location of US Cities')

% Select some cities to display text labels
ind = [6,10,12,16,24,26,53];

% The City object is imported with the data.
text(X(ind,1),X(ind,2),City(ind))
```

Next, we want to find the distance between our data points, which are the locations of the cities in this example. It makes sense to use the Euclidean distance with these data, even though the points are on a sphere. We use the following function call to get interpoint distances. This vector of pairwise distances will be used in the next example, where we illustrate classical multidimensional scaling.

```
% Find the distances between all points.
Dv = pdist(X,'euclidean');
```

7.2.2 Classical MDS

We stated earlier that there are two main categories of multidimensional scaling, one of which is called metric MDS. Classical multidimensional scaling is a special case of metric MDS. We introduce metric MDS here and provide more details in the next section.

We will now focus on dissimilarities as our proximity measures, and these will be denoted by δ_{ij}. The dissimilarities are obtained using the data in the p-dimensional space. We seek a configuration in a lower-dimensional space of size d, such that the distances d_{ij} between the observations in that space are approximately preserved in some sense.

Metric MDS assumes that the dissimilarities and the distances are related in the following manner

$$d_{ij} \approx f(\delta_{ij}) .$$

The function f is a continuous parametric monotonic function [Cox and Cox, 2001], and its form determines the MDS model.

In **classical MDS**, the function f relating distances and dissimilarities is given by the identity function, and we have

$$d_{ij} = \delta_{ij} .$$

If the dissimilarity and the distance are both Euclidean (or assumed to be so), then there is a closed-form solution to find the configuration of points in the lower-dimensional space. Gower [1966] showed the importance of classical MDS, and he called it **principal coordinates analysis**.

We do not go into details on the derivation of classical MDS, but the interested reader can consult the books by Cox and Cox [2001] or Borg and Goenen [2005] for more information. One interesting thing to note is that there are some similarities between PCA and classical MDS. They provide equivalent results when the input dissimilarities are Euclidean distances. This has some implications for situations where one might have a data set with the number of observations n being less than the number of variables p. The covariance matrix is singular in this situation, and one cannot do PCA in such a case. One can always get the distances, which means that we could get something similar to PCA using classical MDS in these cases.

In the next example, we look at how we can get a configuration of points in a d-dimensional space using classical MDS in MATLAB, and we will use the distances that we calculated previously using **pdist**.

Example of Classical MDS

We know that the data were originally in two dimensions on a sphere, given by latitude and longitude. Let's pretend that we do not know anything about the original data. For instance, say we do not know how many characteristics or variables p they represent nor do we know their actual values. All we have are the interpoint distances, and we would like to construct a scatter plot of the data. We can use classical MDS to do this.

The function we will use is called **cmdscale**, and it takes either a full square interpoint dissimilarity matrix as input or the vector of distances one gets from **pdist**. The most general call to **cmdscale** is shown here

$$[\text{Xd,Eig}] = \text{cmdscale(Delta,d)}$$

The second input is optional, and it allows the user to specify the number of dimensions to use for the lower-dimensional space. The default is to use the smallest value of d needed to embed the points. This is determined by the eigenvalues of the matrix derived in the classical MDS methodology.

The first output from **cmdscale** is the configuration or coordinates of the data points in the new space, denoted as X_d. The second output **Eig** (optional) contains the eigenvalues of $X_d \times X_d^T$. We will use **cmdscale** with our interpoint distances from the previous section. Note that we are asking for a configuration with two dimensions or coordinates in the function call below.

```
% We want a 2-D embedding or configuration.
Y = cmdscale(Dv,2);

% Construct a scatter plot.
figure,plot(Y(:,1),Y(:,2),'.')
xlabel('C-MDS Dimension 1')
ylabel('C-MDS Dimension 2')
title('Classical MDS Configuration for US Cities')

% Need to rotate to match the orientation.
view([180,90])
text(Y(ind,1),Y(ind,2),City(ind))
```

The scatter plot of the data in the configuration given by MDS is shown in the bottom of Figure 7.5. We see that the MDS embedding of the points preserves the proximity relationships that we had in the original data. So, points that are close in the original data are also close in the MDS space.

7.2.3 Metric MDS

Metric MDS assumes that the dissimilarities δ_{ij} in the p-dimensional space are related to the distances d_{ij} in d dimensions by a monotonic increasing function f. Classical multidimensional scaling used the identity function to relate them, imposing the requirement that they be equal. There is a closed-form solution for the locations in the transformed space with classical MDS, but other types of metric MDS use the optimization of an objective function to find the configuration of points.

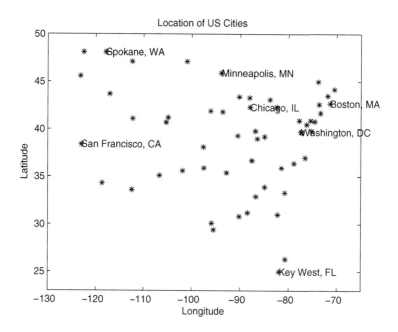

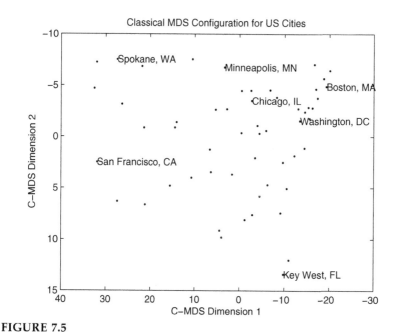

FIGURE 7.5

The first scatter plot shows the location of 56 cities based on their latitude and longitude. The second plot is the configuration from classical MDS. It reconstructs the original data, showing that it preserves the dissimilarities (as seen at the top) in the new MDS space.

The metric MDS implementation in MATLAB distinguishes the various types of metric scaling by means of the different types of objective functions that can be specified. Two of the most common metric scaling types are based on a quantity called *stress*, which is a function of the squared *discrepancies*: $[f(\delta_{ij}) - d_{ij}]^2$. The stress is given by

$$S(d_{ij}) = \sqrt{\frac{\sum_i \sum_j [f(\delta_{ij}) - d_{ij}]^2}{\text{scale factor}}}.$$

Different forms of the *scale factor* in the denominator produce various stress criteria and possibly an alternative configuration of points in the new space.

The MATLAB function for metric MDS is called **mdscale**. There is an input name-value pair that you use to specify the type of scaling. The name of the parameter is called **'criterion'**. We list two popular choices of scaling types here.

- **'metricstress'**: This uses a scale factor given by the sum of squares of the input dissimilarities.

- **'metricsstress'**: This has a scale factor that is the squared stress or the sum of the 4th powers of the input dissimilarities.

There is another option for **'criterion'** called **'strain'**. The user can input this value in **mdscale** to yield something equivalent to classical MDS.

The basic syntax for **mdscale** is

```
Xd = mdscale(Delta,d,'Name','value')
```

The first input **Delta** to **mdscale** is the dissimilarities. These can either be in the vector form from **pdist**, a square matrix of dissimilarities, or a full matrix with similarities. If it has similarities (a symmetric matrix with ones along the diagonal), then MATLAB converts them to dissimilarities. We have to specify the requested number of dimensions *d* with **mdscale**. Note that this is different from **cmdscale** where the default is to find an optimal *d*.

Example of Metric MDS

We are going to use Fisher's iris data to demonstrate the use of **mdscale**. First, we load the data into the workspace, and we then create a data matrix **X**. We need the dissimilarities for MDS, and we will use the **pdist** function to get them. We use the city block metric in this example.

```
load iris
% Put the three variables into one data matrix.
X = [setosa; versicolor; virginica];
```

```
% We need to get the dissimilarities for inputs.
% Let's use the City Block distance as a change.
Dv = pdist(X,'cityblock');
```

We obtain the coordinates of our observations in a 2-dimensional space using this call to **mdscale** and the **'metricsstress'** criterion.

```
% We are ready for metric MDS.
% We will use the 'metricsstress' criterion.
Xd = mdscale(Dv,2,'criterion','metricsstress');
```

We can plot the data in a scatter plot, as shown in Figure 7.6.

```
% First create a vector of group IDs.
G = [ones(1,50),2*ones(1,50),3*ones(1,50)];

% Now get the grouped scatter plot.
% This is in the Statistics Toolbox.
gscatter(Xd(:,1),Xd(:,2),G,[],'.od')
box on
xlabel('Metric MDS Coordinate 1')
ylabel('Metric MDS Coordinate 2')
title('Embedding of Iris Data Using Metric MDS')
```

The configuration of points in a 2-dimensional space is somewhat similar to the results we had from PCA in Figure 7.2.

There is a second output we can get from **mdscale** when we use one of the metric MDS output arguments. This is the observed value of the stress for the given configuration, and it is a measure of how well the distances between the points in the new space approximate the original dissimilarities. Smaller values of the stress criterion are desirable. The following syntax would be used to get the value of stress for the configuration shown in Figure 7.6.

```
[Xd,stress]=mdscale(Dv,2,'criterion','metricsstress');
```

We get a value of 0.038 for the stress.

7.2.4 Nonmetric MDS

The idea behind nonmetric MDS is to relax the assumptions about the function f that relates the input dissimilarities and the distances in the output space. In *nonmetric MDS*, we preserve the rank order of the dissimilarities only, as shown here:

$$\delta_{ij} < \delta_{ab} \Rightarrow f(\delta_{ij}) < f(\delta_{ab}).$$

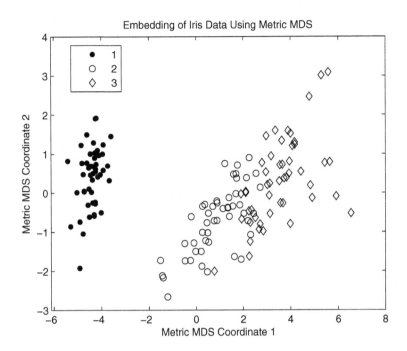

FIGURE 7.6

This is a scatter plot of Fisher's iris data embedded in a 2-dimensional space obtained through metric multidimensional scaling. We used the city block distance and the squared stress criterion with the **mdscale** *function.*

This is called the ***monotonicity constraint***. Essentially, we are seeking an embedding that keeps the ordering of the data, as given by the proximities. The nonmetric MDS approach is also known as ***ordinal MDS*** in the literature.

Because this relaxes some of the constraints regarding the distances between embedded observations, we have to introduce some other notation and terminology. The ***disparity*** measures how well the distance d_{ij} (in the lower d-dimensional space) matches the dissimilarity between objects δ_{ij} (in the higher p-dimensional space). We will denote the disparity by $\hat{d}_{ij}$, and we can now express the monotonicity constraint in terms of the disparities, as shown here:

$$\delta_{ij} < \delta_{ab} \Rightarrow \hat{d}_{ij} \leq \hat{d}_{ab}.$$

Thus, the order of the input dissimilarities between objects is preserved by the disparities. It is interesting to note that ties in the disparities can occur.

Nonmetric MDS algorithms find an embedding in a d-dimensional space by minimizing a stress function—similar to metric MDS. The same **mdscale** function is used to get a configuration of points based on nonmetric MDS. We just specify one of two values for the **'criterion'** parameter to perform nonmetric MDS. These are listed below.

- **'stress'**: This uses a scale factor for stress given by the sum of squares of the input dissimilarities.
- **'sstress'**: This is the squared stress, which has a scale factor that is the sum of the 4th powers of the input dissimilarities.

The stress function in nonmetric MDS is written in terms of the disparities, as shown here

$$S = \sqrt{\frac{\sum_{i<j}(d_{ij} - \hat{d}_{ij})^2}{\text{scale factor}}}.$$

This is sometimes called a *loss function*.

Example Nonmetric MDS

We are going to introduce some different data sets for the next examples. The data set [Wilkinson, 1990] in our first example is discussed in Borg and Groenen [2005]. It is in the form of correlations of crime rates for the 50 states in the US and are based on the Census Bureau's 1970 US Statistical Abstract. One goal might be to use these data to understand the relationship between the crime rates. For example, are the correlations for particular crimes similar? If so, then this could indicate that they happen together.

TABLE 7.2

Correlations of Crime Rates for the US (1970)

Murder	1.00	0.52	0.34	0.81	0.28	0.06	0.11
Rape	0.52	1.00	0.55	0.70	0.68	0.60	0.44
Robbery	0.34	0.55	1.00	0.56	0.62	0.44	0.62
Assault	0.81	0.70	0.56	1.00	0.52	0.32	0.33
Burglary	0.28	0.68	0.62	0.52	1.00	0.80	0.70
Larceny	0.06	0.60	0.44	0.32	0.80	1.00	0.55
Auto Theft	0.11	0.44	0.62	0.33	0.70	0.55	1.00

The correlations for seven crimes are listed in Table 7.2. It is difficult to get an understanding of any patterns by looking at a table of numbers. We can use MDS to find a 2-dimensional representation of the correlations to help us see what is happening. These correlations have been saved in a file called **crime.mat**, along with a variable of labels denoting the type of crime.

```
% The correlations of Crime Rates are from Borg
% and Groenen [2005]. They are saved in crime.mat.
load crime
```

This is a case where we have similarities rather than dissimilarities. In other words, large values of the proximity measure indicate closeness, while small values mean they are different. We can use similarities with the **mdscale** function, rather than dissimilarities, as shown here.

```
% Now get a configuration of points in two
% dimensions using nonmetric MDS.
[Xd,stress] = mdscale(crime,2,'criterion','stress');
```

The following steps produce a scatter plot of the correlations embedded in a 2-dimensional space, which is shown in Figure 7.7. The meaning of the plot is as follows. If two types of crimes (points on the plot) are close, then the corresponding rates are highly correlated. Similarly, if they are not near each other in the plot, then they are not as highly correlated.

```
% Plot the points in 2-D.
plot(Xd(:,1),Xd(:,2),'o')
xlabel('MDS 1')
ylabel('MDS 2')
title('Nonmetric MDS of Crime Rates')
text(Xd(:,1)+0.025,Xd(:,2),crim_lab)
axis([-0.4 .8 -0.3 0.4])
```

We can see some interesting patterns or clusters in the scatter plot that we cannot readily perceive from the table of correlations. For example, assault and murder are close, meaning that they are highly correlated. Furthermore, these are both far from the robbery and auto theft crimes. This makes sense because robbery/theft and assault/murder are different types of crimes. Also, moving from left to right, we go from property crimes to ones that injure people, like rape and murder.

Example—Using Disparities

All we had in this last example was a measure of similarity between our objects. We did not have any actual data to visualize. This is also the case with the second data set that also came from Borg and Groenen [2005]. These data represent similarity ratings for different countries on a scale ranging from 1 to 9, with 1 = *very different* and 9 = *very similar*. The data originally came from Wish [1971], who asked 18 students to rate the global similarity between pairs of countries.

We saved these similarities in a square matrix form in the MATLAB file **countries.mat**. Loading this file will import three variables into the workspace—**nations, nat,** and **nat_lab**. The first contains the original similarities on the 1 to 9 scale. The object **nat** has the entries that have been

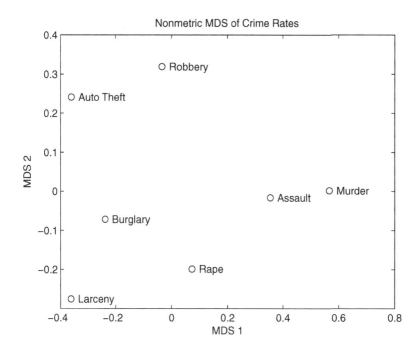

FIGURE 7.7
This is a plot of the crime rate correlations of Table 7.2, where the position of a point is determined by nonmetric MDS. Points that are close together indicate crimes that are highly correlated. We see some distinct groups or clusters in this embedding, which can aid our understanding of the correlations.

transformed to valid MATLAB similarities. The third variable **nat_lab** has labels for the twelve countries.

We will also ask for the additional output from the **mdscale** function that has the disparities. We will use these to construct a diagnostic plot called the Shepard diagram.

```
load countries
% Perform nonmetric MDS using the squared stress.
% We also get the disparities as output.
[Xd,stress,dispar] = ...
    mdscale(nat,2,'criterion','sstress');
plot(Xd(:,1),Xd(:,2),'.')
text(Xd(:,1)+0.01,Xd(:,2)+.02,nat_lab)
title('Nonmetric MDS of Nation Similarities (1971)')
xlabel('MDS 1')
ylabel('MDS 2')
```

The scatter plot is shown in Figure 7.8, and we can see some reasonable locations for countries. The communist countries (China, USSR, and

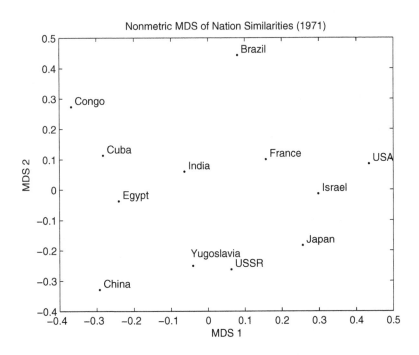

FIGURE 7.8

This shows the 2-dimensional nonmetric MDS embedding for the similarities of countries in 1971. We can see some interesting relationships in these locations. For example, the USSR and Yugoslavia are close together, and China is also located near them. These form a group of communist nations. More developed countries are located in the lower right side of the plot.

Yugoslavia) are located in the lower left quadrant of the plot. Developed nations are on the right side.

We now use the disparities that we obtained from **mdscale** to create a plot that can be used for diagnostics. This is called the **Shepard diagram** and is a scatter plot that shows two sets of points. Normally, the input proximities used as input to MDS are given on the horizontal axis. One set of points in the plot corresponds to the disparities plotted against the inputs. The other set of points shows the distances d_{ij} between the embedded observations versus the input proximities. This will be measured as a Euclidean distance.

We construct this plot for the **countries** data using the following code. It is important to note that the input similarities were converted to dissimilarities by **mdscale**, but we have to explicitly do this to create the plot. This step can be skipped when dissimilarities are used for MDS.

```
% Construct a Shepard diagram.
% First get the Euclidean distances in the
```

```
% MDS space.
dist = pdist(Xd);

% Get unique values below the diagonal of
% disparities matrix.
dispar = squareform(dispar);

% Need to convert input to dissimilarities.
natD = sqrt(1 - nat);

% Converts to vector form.
natV = squareform(natD);

% Now create the plot.
[dum,ord] = sortrows([dispar(:) natV(:)]);
plot(natV,dist,'x',...
    natV(ord),dispar(ord),'*-')
xlabel('Input Dissimilarities')
ylabel('Distances and Disparities')
legend({'Distances','Disparities'})
title('Shepard Diagram')
```

The plot is shown in Figure 7.9. The disparities are the connected points, and we see that they are all on a monotonically increasing line, as required by nonmetric MDS. The x's are the distances in the MDS space, and the vertical distance between the x's and the line show the error in representing that input proximity. We can think of this as a type of residual.

The Shepard diagram provides a visual summary of the scatter around the function that represents the proximities. The errors are used to calculate the stress function. Thus, we can get a sense of which proximity pairs produce large errors (large distance to the line). These are important, because they could be potential outliers.

7.3 Visualization in Higher Dimensions

We already provided some methods for visualizing data in Chapters 2 and 3. However, these were mostly for two dimensions, as we have with the scatter plot. We now present some options for visualizing data in more than two dimensions. We discuss the scatter plot matrix first, and then we describe two ways that we can visualize our data as curves rather than points.

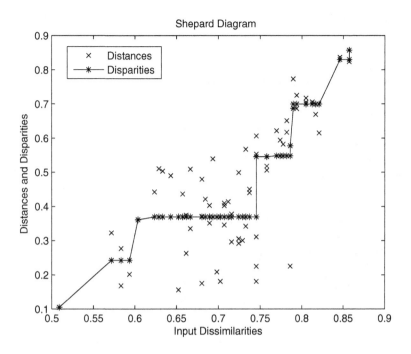

FIGURE 7.9
Here is the Shepard diagram for the MDS embedding of Figure 7.8. The vertical distance between the x's and the line represent the error in representing the proximities.

<u>*TIP*</u>

The orientation of the axes that one gets from MDS is arbitrary, and there is no strict interpretation or meaning that one can ascribe to them. Understanding what they might mean requires knowledge of the subject area. As a consequence, one can reverse the axes or rotate them, as we did with our classical MDS example.

7.3.1 Scatter Plot Matrix

A scatter plot matrix is suitable for multivariate data, where we have $p > 2$. This is essentially a series of 2-dimensional scatter plots that are laid out in a matrix-like format. A *scatter plot matrix* contains all pair-wise scatter plots of the data, and it provides a way to see these relationships.

Each axis of a scatter plot is given by one of the variables. Let the ij-th plot be the one shown in the i-th row and the j-th column. Then the ij-th scatter plot shows the i-th variable on the vertical axis and the j-th variable on the horizontal axis. Therefore, the entries in the upper triangular part of the matrix and the lower part are equivalent. The axes are just transposed. For this reason, some software implementations of a scatter plot matrix provide only the plots in the upper or lower triangular part of the matrix.

The implementation in base MATLAB is called **plotmatrix**, and it gives the entire scatter plot matrix. We find it most useful in our data analysis tasks to call **plotmatrix** with only one argument—the data matrix X. In this case, we would get a $p \times p$ matrix of plots, as we just described. It also provides a histogram of each variable or column of the data matrix along the diagonal matrix of plots.

You can also use two inputs X and Y with the **plotmatrix** function, and MATLAB will plot the columns of Y versus the columns of X. If X has q columns, and Y has r columns, then the resulting scatter plot matrix will be $r \times q$. Thus, it produces a matrix of plots with r rows and q columns.

Example of a Scatter Plot Matrix

We create a scatter plot of Fisher's iris data and display the results in Figure 7.10.

```
load iris
% Put the three variables into one data matrix.
X = [setosa; versicolor; virginica];

% Create a scatter plot matrix
plotmatrix(X)
```

Note that we have a histogram of each variable—sepal length/width and petal length/width. The histograms show the distribution of the individual variables.

Example of a Matrix of Plots

Suppose we had a response variable and predictors. Then, we could use **plotmatrix** to get a matrix of plots that show how the response relates to each of the predictors. For example, we could get a plot of the temperature of the US cities, as follows. The result is shown in Figure 7.11.

```
% Create a plot matrix of January temperature
% against Latitude and Longitude.
load ustemps
dat = [Lat,Long];
plotmatrix(dat,JanTemp)
```

Note that we used the predictors of latitude and longitude as columns in the first input matrix and the vector of responses—temperature—as the second input.

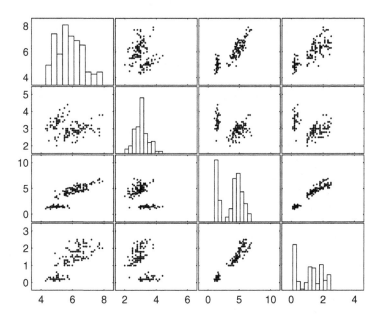

FIGURE 7.10
In this figure, we have a scatter plot matrix of all three species of Fisher's iris data. We see evidence of two groups, but we cannot distinguish the other two species in the larger group.

There is an enhanced version of the scatter plot matrix function that is part of the Statistics Toolbox. This is called **gplotmatrix**. The syntax to create a grouped version of a data matrix is

```
gplotmatrix(X,[],g)
```

where **X** is the data matrix, and **g** is a grouping variable. We include empty brackets for the second input to let MATLAB know that we want a set of plots for the first matrix only. We could specify an additional matrix, in which case it produces a scatter plot matrix based on the columns of the two matrices, as we discussed previously.

Example of a Grouped Scatter Plot Matrix

We illustrate the use of **gplotmatrix** with the **iris** data in our principal component (PCA) space. We repeat the steps to get the principal component scores for ease of reading.

```
load iris
% Put into one matrix.
X = [setosa; versicolor; virginica];
```

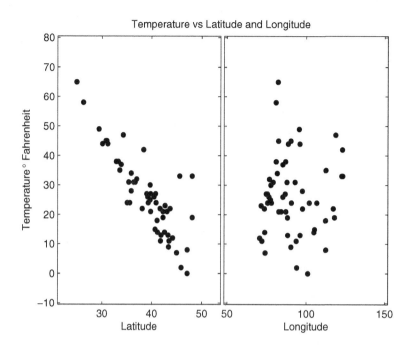

FIGURE 7.11

*This is an example of calling **plotmatrix** with two inputs. Here, we used the predictors of latitude and longitude as columns in the first input and the vector of observed responses in the second.*

```
% Use the PCA function that takes raw data.
[A_pca,Z_pca,vl_pca] = pca(X);
```

The scores are in the output **Z_pca**, which we will use with **gplotmatrix**. We also need a grouping variable and variable names, so we create them first.

```
% Create a grouping variable and a vector of names.
g = [ones(50,1);2*ones(50,1);3*ones(50,1)];
PCNames = {'PC_1','PC_2','PC_3','PC_4'}
gplotmatrix(Z_pca,[],g,[],'.ox',4,...
    'off','variable',PCNames)
title('Fisher''s Iris Data in Principal Components')
```

The grouped plot matrix is shown in the top of Figure 7.12. We specified different marker symbols to make it easier to distinguish the groups in black and white. We have a sense that there is some separability between groups in the first two principal components. Figure 7.12 also has a grouped scatter plot matrix of the data in the original space. We can compare the plots in the two coordinate systems to get an idea of which space would be more useful for subsequent analysis.

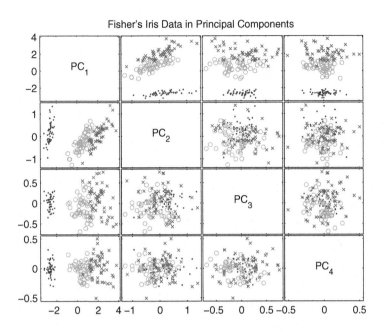

Fisher's Iris Data in Principal Components

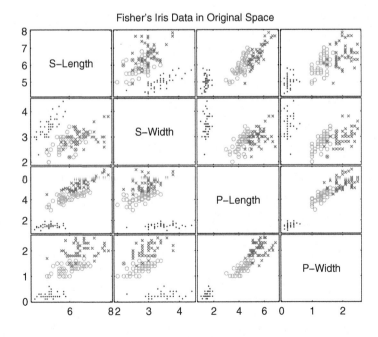

Fisher's Iris Data in Original Space

FIGURE 7.12
These are scatter plot matrices with the points plotted according to their class label.

7.3.2 Parallel Coordinate Plots

One of the problems with a scatter plot matrix is that we see relationships between pairs of variables only, and it is hard to generalize to higher dimensions. In the next two sections, we present methods that allow visualization of all the variables simultaneously. These methods also have some shortcomings that we will point out, but they can be an efficient way to represent many variables.

The first approach is called *parallel coordinate plots*. The Cartesian coordinate system is based on orthogonal axes. We can view a maximum of three dimensions that are projected onto two dimensions, such as a computer screen or a piece of paper. If we relax the orthogonality constraint and make the axes parallel, then we can view many axes on a 2-dimensional display. Parallel coordinate plots were developed by Wegman [1986] and Inselberg [1985].

A parallel coordinate plot is constructed by drawing p lines parallel to each other. Each of the p lines is a copy of the real line representing the coordinate axes for $x_1, x_2, ..., x_p$. The lines are placed an equal distance apart and are usually placed perpendicular to the Cartesian x axis. This produces a horizontal stream of parallel coordinates. Some implementations of parallel coordinate plots place them perpendicular to the y axis, yielding a vertical stack of parallel coordinate axes.

Each value of a variable for an individual observation is plotted as a point on the corresponding parallel coordinate axis. The points are then connected to produce a line. We show a simple example in Figure 7.13 for the vector $(1, 3, 7, 2)^T$.

The Statistics Toolbox has a function for constructing parallel coordinate plots. There are several options that one can use, but the basic syntax is

```
parallelcoords(X)
```

One of the options is to standardize the data first. This can be done based on their eigenvalues or their z-scores (centering at the mean and dividing by the standard deviation). You can also plot them based on groups, as we did with the scatter plot matrix. One useful option with very large data sets is to plot only the specified quantiles.

Example of a Parallel Coordinate Plot

The following creates a parallel coordinate plot for the Fisher's iris data. It is shown in Figure 7.14.

```
% Show the Fisher's iris data in parallel coordinates.
% Load data and put into one matrix.
load iris
X = [setosa; versicolor; virginica];
parallelcoords(X)
box on
```

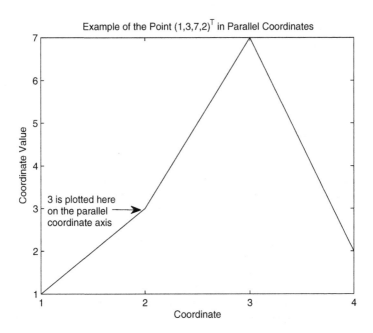

FIGURE 7.13
An example of one 4-dimensional data point displayed in parallel coordinates. The point has the values (1, 3, 7, 2)^T.

```
title(...
     'Parallel Coordinate Plot of Fisher''s Iris Data')
```

One thing we can look for in parallel coordinate plots is evidence of clusters or groups. These are shown as gaps between bundles of lines. We see some indication of two groups in the third parallel coordinate.

One of the problems with parallel coordinates is that we are still viewing the relationship between pairs of parallel coordinates or pairs of variables. Therefore, it is a good idea to re-order the variables (columns of the data matrix), and then reconstruct the parallel coordinate plot to see if other patterns or structure might be visible. This is similar issue to our next topic—Andrews' curves.

7.3.3 Andrews Curves

Andrews [1972] developed a method for visualizing multi-dimensional data by converting each observation into a function. The function used is based on sines and cosines and is given by

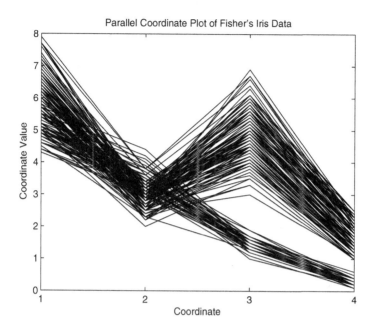

FIGURE 7.14
Here we have a parallel coordinate plot for Fisher's iris data. Note that there is evidence of two groups when we look at parallel coordinate 3, where we see a gap between two bundles of lines.

$$f_{\mathbf{X}}(t) = x_1/\sqrt{2} + x_2\sin(t) + x_3\cos(t) + x_4\sin(2t) + x_5\cos(2t) + \ldots,$$

where $\mathbf{x} = (x_1, x_2, \ldots, x_p)^T$ is one p-dimensional data point. The function is plotted over the range of t, which is $-\pi \leq t \leq \pi$. We see from this expression that the variable values for a given data point are the coefficients of the terms in the Andrews function. We include a subscript $\mathbf{x}$ on $f_{\mathbf{x}}(t)$ to emphasize that one of these curves is created for each of our n observations $\mathbf{x}$.

Example of Andrews Curves

The Statistics Toolbox has a function for displaying observations as Andrews curves. It is called **andrewsplot**. It has most of the same options as the **parallelcoords** function. For example, you can standardize first, display curves corresponding to specified quantiles, or use different colors for groups. This code creates the Andrews curve plot shown in Figure 7.15.

```
% Show Fisher's iris data as Andrews curves
andrewsplot(X)
box on
```

```
title('Andrews Plot of Fisher''s Iris Data')
```

We can look for evidence of groups in these plots, as we did before. There does appear to be two bundles of curves, indicating the presence of at least two groups. However, we still do not see evidence of the three groups known to be in the iris data set.

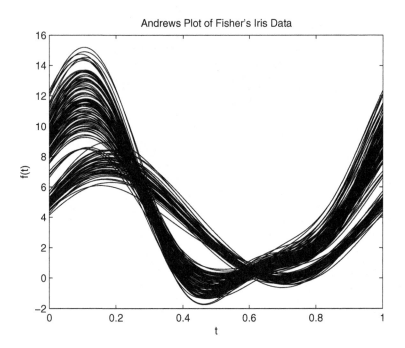

FIGURE 7.15

Each observation in the iris data set is displayed as a curve. There is evidence of at least two groups.

The resulting Andrews curve is dependent on the order of the variables. Those variables that are used as coefficients in the earlier terms have more effect on the shape of the curve. Therefore, it is a good idea to re-order the variables and create another Andrews plot to see what other structure might be discovered.

TIP

If one has a small data set with many variables, then another option is a glyph plot. MATLAB provides a function called **glyphplot** in the Statistics Tool-box. This will create Chernoff faces or star plots. Each observation is displayed as one of these glyphs.

7.4 Summary and Further Reading

In this chapter, we discussed ways to deal with multivariate data. First, we gave some background on MATLAB functions available for transforming data to different coordinates using principal component analysis (PCA). There is an additional PCA function in base MATLAB called **eigs**, which operates on large, sparse matrices. One nice aspect of this function is that it allows the user to ask for only d eigenvectors and eigenvalues, which saves computational time and memory.

A methodology for data transformation we did not discuss is called *nonnegative matrix factorization* (NMF). There is a function in the Statistics Toolbox called **nnmf** that will return this data transformation. NMF seeks to decompose a matrix into a product of two matrices with all nonnegative entries. This approach is particularly suitable in situations where the original matrix is nonnegative. The factored matrices are easier to interpret in this situation, which is often not the case with PCA.

The Statistics Toolbox has a function that will perform probabilistic PCA [Tipping and Bishop, 1999] that is sometimes better to use when one has missing data. The MATLAB function for this is called **ppca**.

We next presented several multidimensional scaling methods for creating configurations of observations in a lower-dimensional space. The benefit of these methods is that they can be used in cases where we have information only about how similar or dissimilar observations are from each other.

We concluded the chapter with a discussion of ways we can visualize data in more than two dimensions. This included scatter plot matrices, parallel coordinate plots, and Andrews curves. The MATLAB functions we discussed in this chapter are summarized in Tables 7.3 and 7.4.

There are several excellent books on multidimensional scaling. The first is a very readable book by Cox and Cox [2001]. Another is the text by Borg and Groenen [2005]. We also suggest the treatment written by Kruskal and Wish [1978]. For principal component analysis, we highly recommend Jackson [1991] and Joliffe [1986].

There are not a lot of books available just on visualizing high-dimensional data. One that we can recommend is the *Graphics of Large Datasets: Visualizing a Million* [Unwin et al. 2006]. Theus and Urbanek [2009] have a one on interactive graphics and how they can be used to analyze high-dimensional data. Another excellent book is by Cook and Swayne [2007].

There is a chapter called *Multivariate Data Analysis* in the MATLAB online documentation. It has several sections that are relevant for the topics we just discussed. These are *Multidimensional Scaling, Principal Component Analysis, Nonnegative Matrix Factorization*, and *Multivariate Visualization*. Finally, the *Mathematics* chapter in the base MATLAB documentation has a section called *Linear Algebra* that covers eigenvectors and eigenvalues.

TABLE 7.3

Functions for Transforming Data

Function Name	Description
cmdscale	Classical multidimensional scaling
eig	Eigenanalysis in base MATLAB
eigs	Eigenanalysis for sparse matrices
mdscale	Metric and nonmetric multidimensional scaling
nnmf	Nonnegative matrix factorization (NMF)
pca	Principal component analysis (PCA) on raw data
pcacov	PCA with covariance or correlation matrix

TABLE 7.4

Functions for Data Visualization

Function Name	Description
andrewsplot	Andrews plots
biplot	Biplot of data transformed with PCA or NMF
glyphplot	Star plots or Chernoff faces
gplotmatrix	Grouped scatter plot matrix
parallelcoords	Parallel coordinates plot
plotmatrix	Matrix of scatter plots

Chapter 8

Classification and Clustering

The Milky Way is nothing else but a mass of innumerable stars planted together in clusters.
  Galileo Galilei (1564–1642)

In this chapter, we are going to cover two of the main areas or tasks in the area of machine learning. The first is referred to as classification, pattern recognition, or supervised learning. The second one is an exploratory data analysis technique known as unsupervised learning or clustering. We will briefly describe some background for the methods and then demonstrate the MATLAB® functions. All of the functions for supervised and unsupervised learning described in this chapter are from the Statistics Toolbox.

8.1 Supervised Learning or Classification

In *supervised learning*, we have a data set with a known number of classes or groups. Furthermore, each of our observations has a class label attached to it. This is what we have with Fisher's iris data. There are three known groups or species of iris, and each of our observations has the given species name associated with it. Our goal is to develop or build a *classifier* using the data. We then use the classifier to predict the class membership or label of future observations.

This type of situation arises in many applications. For example, a doctor might use a classifier to assign an illness based on patient symptoms or test results, a market manager has to determine what people should receive their advertisements based on demographic information, or an ecologist needs to classify regions in an image as water or land.

These examples have several things in common, and we point them out in the following list of general steps one takes in classification tasks.

1. Determine what features will be collected. These features are the same thing as our variables in previous chapters. The features in the iris data are the widths and lengths of the petals and sepals. In the advertisement example, the features might be gender, age, area of residence, and so on. This is a key first step because building a classifier is much easier with features that will distinguish the classes.

2. Once we have the features selected, we then measure n_m of them for each of M groups. With the iris data, we have $M = 3$ groups, and $n_m = 50$, $m = 1, ..., 3$. This is sometimes called the *training set*.

3. We are now ready to build or estimate a classifier using the observations with the known true class labels. This process is sometimes referred to as *training the classifier*.

4. The last step is to assess the performance of the classifier. We could use percent correctly classified, confusion matrices, or performance curves for this purpose.

The last two steps of this process will be the main focus of this section. We would like to point out that the process of building a classifier in supervised learning is similar in some respects to model building and regression analysis (Chapter 6). In classification, the *response* variable is a *class* label, and our *predictor variables* are often called *features*. We do not cover it in this book, but there are also regression methods that can be used in a classification problem, when we think of the application in this manner [Bishop, 2006].

MATLAB categorizes supervised learning methods into two main types—parametric and nonparametric. These have a meaning similar to what we had in regression analysis. A parametric classification method uses a parametric model that is estimated using the data, while nonparametric approaches do not.

In this section, we describe two parametric methods, which are called naive Bayes and discriminant analysis in the MATLAB documentation. We will also present some ways to evaluate the performance of our classifier. After this, we discuss a popular nonparametric approach called k-nearest neighbors. However, we first present some background information on the general Bayesian approach to classification.

8.1.1 Bayes Decision Theory

In Bayes decision theory, the classification problem is specified in terms of probabilities. Probabilities are usually easy for people to understand, and they provide a nice way to assess the performance of the classifier. We will see that to build or train a classifier, we are essentially estimating probability density functions from the data.

We will start our discussion by setting some notation and defining the probabilities we need. The class membership will be represented by ω_m, with $m = 1, ..., M$. The features (or variables) we use to build the classifier will be denoted by our p-dimensional observation vector $\mathbf{x}_i$. Recall that in supervised learning, we also have an additional variable associated with each observation, which corresponds to the true class label.

The **posterior probability** is the probability that an observation $\mathbf{x}_i$ belongs to the m-th class, as shown here

$$P(\omega_m | \mathbf{x}_i); \quad m = 1, ..., M.$$

The class label is represented by ω_m. It makes intuitive sense to assign an observation to the highest class posterior probability given above. Thus, we need to come up with a way to estimate or fit these probabilities.

We can use Bayes' Theorem to decompose the posterior probability, as follows

$$P(\omega_m | \mathbf{x}_i) = \frac{P(\omega_m) P(\mathbf{x}_i | \omega_m)}{P(\mathbf{x}_i)},$$

where

$$P(\mathbf{x}_i) = \sum_m P(\omega_m) P(\mathbf{x}_i | \omega_m).$$

The probability $P(\omega_m)$ is called the **prior probability**, and $P(\mathbf{x}_i | \omega_m)$ is called the **class-conditional probability**.

The prior probability represents how likely it is that an observation is in that class without knowing anything else about it, such as the features we measured. An example of this might be situations where we know how common a disease is in the population or what percentage of an area is covered by water.

The class-conditional probability is the probability of observing a feature vector $\mathbf{x}_i$, given it is in class ω_m. We can think of this as the probability distribution of the features for each separate class, and each class will have its own class-conditional probability.

To build or train our classifier, we have to estimate the prior and class-conditional probabilities. The prior probability can either be inferred from prior domain knowledge, estimated from the observed data, or assumed to be equal across the classes.

If we estimate these from the data, then the prior probabilities are the relative frequency of the observations we have in each class, as shown here

$$\hat{P}(\omega_m) = \frac{n_m}{n},$$

where n_m is the number of observations in the m-th class. Using this with the Fisher's iris data, we would have equal priors given by $1/3$.

Estimating the class-conditional probabilities is a little more involved, and the various choices give rise to the two parametric approaches in MATLAB. Both involve estimating the class-conditional probabilities for each class, using the observations in the training data set.

8.1.2 Discriminant Analysis

In *discriminant analysis*, one estimates the class-conditional probabilities using normal (or Gaussian) distributions. There are two choices one can make about the distributions, which produce either a linear or quadratic classifier. Our subsequent discussion assumes that we have more than one feature, but the ideas carry over to the univariate case.

If we assume that the covariance matrix for each class is the same, then we will have a *linear classifier*. If this assumption is relaxed, and we allow the covariance matrices to be different, then we get a *quadratic classifier*.

We estimate the class-conditional probabilities $P(\mathbf{x}_i|\omega_m)$ in a linear classifier by first estimating the p-dimensional means for each class, based only on those cases that are in class ω_m. The corresponding class means are subtracted from the observations. The sample covariance is then obtained using all of these centered data.

With quadratic classifiers, the classes are assumed to have different covariance matrices. We first center the data as we did in the linear classifier, but we now estimate the covariance separately for each class.

The linear and quadratic classifiers we build divide the feature space into regions based on the estimated probabilities. If a new observation falls into a region where the posterior probability for class ω_m is highest, then we would predict that any new observation is a member of class ω_m. Linear classifiers produce regions that are separated by hyperplanes. The regions we get from quadratic classifiers are given by hypersurfaces, such as ellipses, hyperbolas, or parabolas.

MATLAB has an object class called a `ClassificationDiscriminant`, and it is created using the function called `fitcdiscr`. The function requires two inputs—a matrix of features or variables **X** and a vector of class labels **y**.

There are many options for the name-value pairs that provide a lot of flexibility for the user. The interested reader should consult the help files for more details. For now, we just mention how we can get a linear or quadratic classifier. The basic syntax for a liner classifier is

```
class_obj = fitcdiscr(X,y,'DiscrimType','linear').
```

A quadratic classifier is obtained using

```
class_obj = fitcdiscr(X,y,'DiscrimType','quadratic').
```

Other options for the function will be explored in the coming examples.

Example—Building a Classifier

We will use Fisher's iris data for all of the examples in this chapter. Recall from Chapter 1 that the original application for these data was to build a classifier that could predict the species (or the classes). Thus, they are particularly suitable for our examples in this chapter. We get things going by loading the data into the workspace and then creating our data matrix with the features and a vector that contains the species or class labels.

```
% Load the iris data. This imports three data objects:
% setosa, versicolor, and virginica.
load iris

% Create a data matrix.
X = [setosa; versicolor; virginica];

% Create a vector of class labels.
y = [ones(50,1); 2*ones(50,1); 3*ones(50,1)];
```

We now create a cell arrays of strings to specify the feature names. This is followed by two calls to the **fitcdiscr** function—one to create a linear classifier and one to create a quadratic classifier. In the code below, we are specifying some of the default inputs for readability. The default classifier is linear, and the default for the prior probability is to estimate it from the data. The other options for the **'Prior'** name argument include uniform priors or a vector of values given by the user.

```
% Create feature names.
ftns = {'S-Length','S-Width','P-Length','P-Width'};

% Create a linear classifier.
% The 'DiscrimType' default is 'linear'.
% The 'Prior' default is 'empirical'.
% We specify them here for clarity.
Lclass = fitcdiscr(X,y,'DiscrimType','linear',...
    'PredictorNames',ftns,...
    'Prior','empirical');

% Create a quadratic classifier.
Qclass = fitcdiscr(X,y,'DiscrimType','quadratic',...
    'PredictorNames',ftns,...
    'Prior','empirical');
```

Example—Evaluation using Resubstitution Error

One way we can evaluate the performance of the classifiers is to find the *resubstitution error*. This is the rate at which observations are misclassified using our classifier. A large value of this error indicates that the classifier is not a good one. This could be due to the features or the type of classifier used. The following MATLAB code finds the resubstitution error for our classifiers.

```
% Calculate and show the error for the linear
% classifier.
Lrerr = resubLoss(Lclass)

Lrerr =

    0.0200

% Calculate and display the error for the quadratic
% classifier.
Qrerr = resubLoss(Qclass)

Qrerr =

    0.0200
```

The resubstitution errors are low, which means that the performance of the classifiers is good. The classifiers have the same on error for this data set. It is usually good practice to keep the simpler model or classifier. So, we should use the linear classifier as our model, based on this assessment.

It should be noted that the resubstitution error is based on predicting observations that were used to build the classifier, so the classifier has already seen them. Therefore, the resubstitution error is biased low and is not necessarily a good indication of how the classifier will perform with new data.

Example—Evaluation using Cross-validation

One way to handle this bias is to train or build the classifier using one data set, and employ a different data set with class labels to test it. However, it is often too expensive to have separate training and testing data sets. This is where cross-validation can help. *Cross-validation* is an approach where the data set is partitioned into K disjoint subsets. The software will loop through the K subsets, leaving out one for testing purposes, and building the classifier using the other $K - 1$. We end up with class predictions for all n data points serving as members of one of the K independent testing sets.

We can obtain the cross-validation prediction error using our classifier data objects. First, we have to create a cross-validation model object using the **crossval** function. The default value for K is 10, but we ask for 5. We get an output from **crossval** that is a special object, which is then used with the function **kfoldLoss** to calculate the error.

```
% Get the cross-validation (CV) prediction error.
% The crossval function creates a CV object.
Lcv_mod = crossval(Lclass,'kfold',5);
Qcv_mod = crossval(Qclass,'kfold',5);

% Now get the error based on CV.
Lcverr = kfoldLoss(Lcv_mod)
Qcverr = kfoldLoss(Qcv_mod)
```

The resulting errors are given here.

```
Lcverr =

    0.0200


Qcverr =

    0.0333
```

The cross-validation error is slightly different for the quadratic classifier, while the error is the same for the linear classifier.

The cross-validation approach to estimating performance can be applied to any model, including ones that we get from regression analysis and other classifiers. MATLAB provides methods that work on certain model data objects, such as one might get from decision trees or regression. Furthermore, other types of errors such as mean square error can be calculated with these methods.

It is possible to get the percentage of the observations that are correctly classified from our prediction error results. Simply subtract the error from one and convert to a percent. For the iris data, our estimated percentage of observations that would be correctly classified is 98% for a linear classifier, which is quite high.

Example of the Confusion Matrix

We can also assess the classifiers by looking at the *confusion matrix*. The *ij*-th element of the confusion matrix represents the number of observations of class *i* that the classifier says belongs to class *j*. In other words, it shows how many classification errors we have, and what classes were confused by the classifier. We get the confusion matrix for the quadratic classifier here.

```
% resubPredict provides the resubstitution response
% of a classifier.
R = confusionmat(Qclass.Y,resubPredict(Qclass))
```

```
% The following is displayed in the window.

R =

    50     0     0
     0    48     2
     0     1    49
```

We see from this result that the confusion matrix is of size $M \times M$. All 50 cases in the *Setosa* species were correctly classified because **R(1,1)** is 50. This is not surprising, given the scatter plots and other high-dimensional plots shown in Chapter 7, where we saw that the *Setosa* observations were quite distinct from the other two groups. There does seem to be some confusion between the *Versicolor* and *Virginica* species, as expected. The element in **R(2,3)** tells us that there were two *Versicolor* cases that were misclassified as *Virginica*. Similarly, the element **R(3,2)** indicates that one *Virginica* iris was classified as *Versicolor*.

The sum of the off-diagonal elements in the confusion matrix yields the total number of misclassified observations. We can also get that value as the product of the classification error and the total number of observations n.

```
% We can get the number of misclassified points for
% the linear classifier by multiplying the error by n.
Lrerr*150

ans =

    3.0000
```

TIP

Our analysis so far has treated the costs of misclassification the same for every observation and class—the cost is 0 for correct classifications and 1 for incorrect classifications. You can create a cost matrix that changes this and use it as an input when creating the classifier.

8.1.3 Naive Bayes Classifiers

The *naive Bayes classifier* estimates the class-conditional probabilities in a different manner. The approach makes the assumption that the individual features are independent, given the class. Therefore, the probability density function for the within-class conditional probability can be written as

$$P(\mathbf{x}|\omega_m) = P(x_1|\omega_m) \times \dots \times P(x_p|\omega_m).$$

In other words, using the data within a class, we first estimate the univariate density for each feature or dimension. We then multiply them together to get the joint density. This can save on computations because fewer parameters need to be estimated. Another benefit of fewer parameters to estimate is that we typically require fewer data points for training in these situations, which can be important when we have many features or dimensions to deal with.

MATLAB has a function for constructing a naive Bayes classifier, which is called **fitNaiveBayes**. It is an object-oriented function that produces a special type of data object called **NaiveBayes**. There are five distributions that can be specified for the class-conditional probabilities. These include the normal distribution, kernel density estimates, multinomials, and strings that name other probability distributions from the Statistics Toolbox. This provides a lot of flexibility for modeling the class-conditional probabilities.

Example of Naive Bayes Classifier

We continue to use the same Fisher's iris data to show how to build a naive Bayes classifier. Recall that the **X** matrix below contains the four features for the three species of iris, and the class labels are 1 (*Setosa*), 2 (*Versicolor*), and 3 (*Virginica*). Our first naive Bayes classifier will use the kernel density method to estimate the individual probabilities for each feature and within each class.

```
% Use iris data and kernel option.
NBk = fitNaiveBayes(X,y,'distribution','kernel')

% This is the resulting classifier object.
NBk =

Naive Bayes classifier with 3 classes for 4 dimensions.
Feature Distribution(s):kernel
Classes:1, 2, 3
```

We asked MATLAB to display the **NaiveBayes** object at the command line, and we see some summary information about the object. We can also check the type or class of the object using this function call.

```
class(NBk)

% This is the class of the object.
ans =

NaiveBayes
```

We can assess the results of our classifier by finding the confusion matrix, as we show here.

```
% Get the confusion matrix for this case.
predNBk = predict(NBk,X);
```

```
confusionmat(y,predNBk)

ans =

    50     0     0
     0    48     2
     0     3    47
```

This is a little worse than the linear and quadratic classifiers. We now have five observations that are incorrectly classified. Therefore, the resubstitution error for this classifier is $5/150 \approx 0.033$.

Let's see what happens if we ask for class-conditional probabilities that are based on the normal distribution. Here is the code.

```
% Also get the Gaussian and compare the error.
% The 'normal' is the default, but we specify it
% for clarity.
NBn = fitNaiveBayes(X,y,'distribution','normal')

% Get the confusion matrix for this case.
predNBn = predict(NBn,X);
confusionmat(y,predNBn)
```

The **NaiveBayes** object and the confusion matrix are shown below.

```
NBn =

Naive Bayes classifier with 3 classes for 4 dimensions.
Feature Distribution(s):normal
Classes:1, 2, 3

ans =
    50     0     0
     0    47     3
     0     3    47
```

Now, we have six observations that are misclassified, and the resubstitution error is even larger: $6/150 = 0.04$.

8.1.4 Nearest Neighbor Classifier

MATLAB provides several nonparametric approaches for classification or supervised learning. These include support vector machines, classification trees, ensemble methods, and nearest neighbors [Bishop, 2006]. The last of these is the topic of this section.

The K-nearest neighbor classifier uses a rule that is based on what class occurs most often in the set of K-nearest neighbors of a feature vector that we want to classify. We can also cast it in the same framework we have been using so far, where we are classifying an observation based on the highest posterior probability $P(\omega_m|\mathbf{x})$, $m = 1, ..., M$.

Suppose we want to classify a feature vector $\mathbf{x}$, which is not necessarily one of our training points. We construct a sphere centered at this data point $\mathbf{x}$, and the size of the sphere is such that exactly K points in our training data set are within that sphere. It does not matter what class they belong to—we just want K-labeled points from our training data in the sphere.

We let K_m represent the number of points in the sphere that are from the m-th class. An estimate of the density associated with each class in the sphere is given by

$$P(\mathbf{x}|\omega_m) = \frac{K_m}{n_m V},$$

where V is the volume of the sphere and n_m is the number of observations in class m. The unconditional density is found using

$$P(\mathbf{x}) = \frac{K}{nV},$$

and the priors are estimated from the training data as

$$P(\omega_m) = \frac{n_m}{n}.$$

Putting these together using Bayes' theorem, we get the following posterior probability:

$$P(\omega_m|\mathbf{x}) = \frac{P(\omega_m)P(\mathbf{x}|\omega_m)}{P(\mathbf{x})} = \frac{K_m}{K}.$$

We would assign the class label m corresponding to the highest posterior probability as given above.

So, to classify a new observation, we first have to specify a value of K. We then find the K closest points to this observation that are in our training data set. We assign the class label that has the highest number of cases in this nearest neighbor set.

MATLAB's function for constructing a nearest neighbor classifier is called **fitcknn**. The general syntax for this function is shown below.

```
mdl = fitcknn(X,y,'name',value)
```

As in our previous examples, the name-value pairs are optional and allow one to specify many aspects of the classification model, including the number of neighbors K. The default is $K = 1$.

Example of K-Nearest Neighbor Classifier

The data object that is created with **fitcknn** is a **ClassificationKNN** object. One can use many of the same methods with this object class that we saw with the linear and quadratic classifier objects. We first fit a K-nearest neighbor classifier, requesting $K = 5$ neighbors.

```
% Use the same iris data from before.
Knn = fitcknn(X,y,'NumNeighbors',5)

% This is a summary of the KNN object.
Knn =

  ClassificationKNN
       PredictorNames: {'x1'   'x2'   'x3'   'x4'}
        ResponseName: 'Y'
          ClassNames: [1 2 3]
      ScoreTransform: 'none'
     NumObservations: 150
            Distance: 'euclidean'
        NumNeighbors: 5
```

The description of the classifier object is shown above, and we see one field has a distance name attached to it. The default is to use Euclidean distance to determine the closest neighbors in the data set. How we measure the distance can have an effect on the classification results. Thus, it might be a good idea to try different distances with a K-nearest neighbor classifier and see if that changes the performance of the classifier.

We can also get both the resubstitution error and the cross-validation error, as we did earlier with the linear and quadratic models.

```
% Get the resubstitution error.
Krerr = resubLoss(Knn)

% Get the cross-validation object first.
Kcv_mod = crossval(Knn,'kfold',5);

% Now get the error based on cross-validation.
Kcverr = kfoldLoss(Kcv_mod)
```

The resulting errors are shown here.

```
Krerr =

    0.0333

Kcverr =

    0.0267
```

We see that the results are in keeping with the discriminant analysis and naive Bayes approaches.

Finally, let's see what the confusion matrix looks like for this *K*-nearest neighbor classifier.

```
% Get the confusion matrix.
R = confusionmat(Knn.Y,resubPredict(Knn))

% The confusion matrix is shown here.
R =

    50    0    0
     0   47    3
     0    2   48
```

This is similar to the confusion matrix for the naive Bayes classifier with the kernel density estimation method. The confusion between *Versicolor* and *Virginica* is different with this classifier.

<u>*TIP*</u>

Try different values for the *K* nearest neighbors. See what happens when you use alternative ways to measure distance. Then compare the errors of the resulting *K*-nearest neighbor classifiers to find a good one.

8.2 Unsupervised Learning or Cluster Analysis

Clustering data is the process of organizing our data into groups, such that observations placed within a group are more similar to each other than they are to observations in other groups. Clustering is also known as *unsupervised learning*. With unsupervised learning tasks, we typically do not have class labels associated with each observation. Or, we could apply clustering analysis when we choose not to use the class labels, as we do in our examples.

There is also an inherent assumption in clustering—that there exist some clusters or groups in the data set. Determining whether or not there are two or more groups in the data set will not be covered in this text [Martinez et al., 2010]. However, we will discuss the issue of estimating the number of groups and assessing the content of the clusters in this primer. Most clustering methods will find some specified number of groups, but what we really desire from this process are clusters that are meaningful and also provide information about the phenomena we are studying.

We will see in the examples that the resulting clusters are highly dependent on the algorithms and distances that are applied to the data. For instance, if our algorithm looks for spherical clusters, then we will get spherical clusters. However, they might not be meaningful nor offer any insights. Clustering data is an exploratory data analysis task. So, the analyst should look at the groups that result from the clustering and determine whether or not they are indicative of true structure in the data. Additionally, one should apply different clustering techniques and examine the groups to better understand the data.

We cover two of the basic clustering approaches in this chapter. First, we provide a discussion of agglomerative clustering, which is a hierarchical approach. This is followed by a popular method called *K*-means, which is particularly appropriate for large data sets.

8.2.1 Hierarchical Clustering

Hierarchical clustering approaches partition data into groups using multiple levels. We can think of it is as a hierarchical tree or a set of nested partitions. The clustering process consists of a series of steps, where the two clusters are either divided or merged at each step.

The first approach is called *divisive clustering*. Clusters are formed based on some optimality criterion or objective function. The method starts by placing all *n* data points in a single cluster or group. A cluster is then split at subsequent stages—again to optimize the criterion. The process stops when there are *n* singleton clusters or until some given number of clusters are formed. Divisive hierarchical clustering methods are less common, and they can be computationally intensive.

MATLAB has implemented the more common *agglomerative clustering* methodology in the Statistics Toolbox. In this type of clustering method, we start off with *n* groups—one for each data point, and at the end of the process, we have just one cluster of size *n*. Thus, we see that the agglomerative and divisive hierarchical approaches address clustering from opposite directions. We now describe the three main components of agglomerative clustering —calculating distances, forming linkages, and choosing the number of clusters.

Distance

We discussed various types of distances in Chapter 7, where we described their use with multidimensional scaling. Measuring the distances between observations also plays a key role in agglomerative clustering. Recall that the MATLAB function for calculating distances is **pdist** and that it returns the elements in the upper-triangular part of the interpoint-distance matrix in a vector form. This function can also be used with the agglomerative clustering process.

Table 7.1 lists the types of distances one could obtain using **pdist**. The distance used can have an impact on the groups that are found. So, it is a good idea to apply agglomerative clustering using different distances and then explore the results.

Linkage

We said that the agglomerative clustering method joins the two closest clusters at each step. It is easy to understand what this means when we have two groups with just one data point each. However, it is more challenging to measure closeness when there are two or more observations in a group. This is where linkage comes in.

Linkage is the method for determining how the proximity relationship between clusters is measured. Let r and s denote two clusters, with n_r and n_s observations, respectively. We denote the distance between the r-th and s-th cluster by $d_c(r, s)$. This distance is used to define the three main linkage options available in MATLAB.

It seems that the default linkage type in most software implementations of agglomerative clustering is *single linkage*, and MATLAB is no exception. It is also known as *nearest neighbor linkage* because the distance between the clusters is given by the smallest distance between pairs of observations—one from each cluster. The linkage distance is given by

$$d_c(r, s) = \min\{d(\mathbf{x}_{ri}, \mathbf{x}_{sj})\} \quad i = 1, ..., n_r \; ; \; j = 1, ..., n_s,$$

where $d(\mathbf{x}_{ri}, \mathbf{x}_{sj})$ is the distance between the i-th observation in cluster r and the j-th observation in cluster s. It is unfortunate that single linkage is the default option because it can produce rather strange cluster structures (called chaining), as we show in the examples.

Complete linkage is the opposite of single linkage. The proximity of groups is measured by the largest distance between data points, with one coming from each group. The distance between clusters using complete linkage is

$$d_c(r, s) = \max\{d(\mathbf{x}_{ri}, \mathbf{x}_{sj})\} \quad i = 1, ..., n_r \; ; \; j = 1, ..., n_s.$$

Complete linkage tends to produce spherical clusters.

Another method is called *average linkage*, where we define the distance between groups as the average distance between pairs of points. This is given by

$$d_c(r, s) = \frac{1}{n_r n_s} \sum_{i=1}^{n_r} \sum_{j=1}^{n_s} d(\mathbf{x}_{ri}, \mathbf{x}_{sj}).$$

Average linkage tends to combine clusters that have small variances. It also tends to yield clusters with approximately equal variances. MATLAB also has a weighted version of this linkage.

As we mentioned previously, these linkage methods can produce very different clusters. When we combine the options for linkage with the many choices for measuring distance between observations, we could get even more cluster structures to explore.

The MATLAB function **linkage** will take the proximity information and use it to group the closest pair of clusters, according to the chosen linkage method. There are two main options for calling this function. One can input the raw data **X** and specify the distance and linkage method. Or, one could provide the vector of interpoint distances from **pdist**. The function returns information to construct a hierarchical tree, which we describe next. Here is an example of **linkage** using the raw data

```
Z = linkage(X,'method','metric'),
```

where **'method'** specifies the linkage to use, and **'metric'** is one of the options used with **pdist**. We illustrate these options in our examples.

Number of Clusters

As shown above, the **linkage** function returns a matrix **Z** that contains information MATLAB needs to create a binary hierarchical tree. This can be shown in a plot called a *dendrogram*. This is a tree diagram that shows the nested structure of the merges and possible groupings (partitions) at each stage.

The dendrogram can be displayed vertically or horizontally, and MATLAB provides options for different orientations. We will focus on the vertical top-to-bottom version, where the root node is at the top. This node represents the cluster comprised of all n data points. The leaf nodes are at the bottom of the tree diagram, and they correspond to the singleton clusters. The default in MATLAB is to show thirty (30) leaf nodes to avoid over-plotting, but one can ask to display more or fewer leaf nodes.

The MATLAB function to use is called **dendrogram**. We generated some data with two clusters, and we show the resulting dendrogram in Figure 8.1.

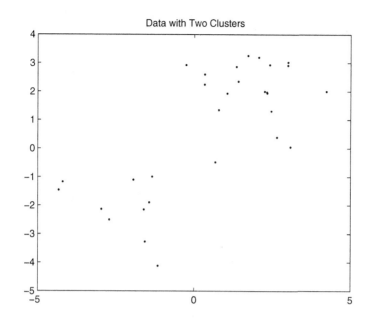

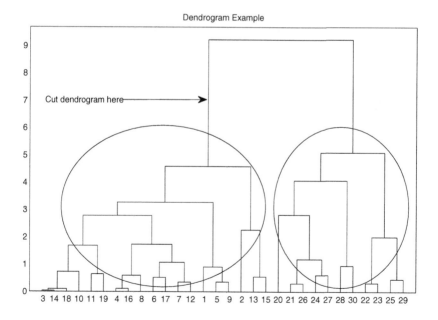

FIGURE 8.1

This is a dendrogram for the data shown in the above scatter plot, where we see evidence of two groups. The dendrogram was created using complete linkage and Euclidean distance. We get two clusters, as shown by the ellipses, if we cut the dendrogram around 7 on the vertical axis.

```
% Create some two-cluster data.
% Use this to create a dendrogram.
X = [randn(20,2)+2; randn(10,2)-2];
plot(X(:,1),X(:,2),'.')

% Get the dendrogram using the raw data,
% complete linkage, and Euclidean distance.
Z = linkage(X,'complete');
dendrogram(Z)
title('Dendrogram Example')
```

We see that the dendrogram is comprised of inverted U-shaped links, where the top of the inverted U shows the place at which two clusters are merged. The vertical axis usually corresponds to the value of the distance between the two clusters.

The heights of the Us represent the distance between the merged groups. If this is large, then we have groups being merged that are far apart. This is evidence that these could be separate clusters. In Figure 8.1, we have the last merge taking place at a value of 9 on the vertical axis. The height of that link is large, which means that we might have two groups in the data set. If we cut the tree at around 7 (on the vertical axis), then we would get two clusters.

The final step is to extract the cluster IDs based on the distance and linkage used in the agglomerative clustering algorithm. The **cluster** function will provide these IDs for a specified number of clusters. It takes the output from **dendrogram**, as shown here

```
cid = cluster(Z,'MaxClust',N)
```

This syntax asks for a given number of **N** clusters. The output vector **cid** is of size n. It has the cluster label for each observation.

There is an additional function for agglomerative clustering that will take care of all three steps—distance, linkage, and cluster labeling. It is called **clusterdata**. There are name-value pairs for distance metrics, linkage methods, and selecting the number of clusters. There is also an option to save memory when working with very large data sets. The following produces the same results we had in Figure 8.1

```
cid = clusterdata(X,'linkage','complete','MaxClust',2)
```

Example of Agglomerative Clustering

We will use the **iris** data to illustrate agglomerative clustering in MATLAB. It turns out that we have class labels for these data, and we also know that there are three groups or clusters, given by the species. However, we will ignore this information and use agglomerative clustering to see if there is some evidence of groups. We first find the Euclidean distance and then apply the default of single linkage.

```
% Load the data.
load iris

% Create a data matrix.
X = [setosa; versicolor; virginica];

% Agglomerative clustering with defaults
% of Euclidean distance and single linkage.
% First, find the Euclidean distance.
Euc_d = pdist(X);

% Get single linkage, which is the default.
Zs = linkage(Euc_d);

% Construct the dendrogram.
dendrogram(Zs)
title('Single Linkage - Iris Data')
```

The dendrogram is shown in the top of Figure 8.2. There is some chaining with this data set, which can be seen as the sequence of merges on the left side of the dendrogram. The single linkage method is merging groups of size one towards the end of the algorithm.

Let's keep the same Euclidean distance, but use complete linkage to merge groups. Here is the code.

```
% Try complete linkage.
Zc = linkage(Euc_d,'complete');
dendrogram(Zc)
title('Complete Linkage - Iris Data')
```

The dendrogram is shown in Figure 8.2 (bottom), and there is no chaining. We also see some evidence that three groups would be a reasonable number to specify by looking at the heights of the inverted Us.

We know that there are three groups in the **iris** data set. We can ask for a partition of three clusters based on single and complete linkage.

```
% Get three clusters from each method.
cidSL = cluster(Zs,'maxclust',3);
cidCL = cluster(Zc,'maxclust',3);
```

The **tabulate** function provides a summary of the cluster IDs.

```
tabulate(cidSL)
% This is displayed.
```

Value	Count	Percent
1	2	1.33%
2	98	65.33%
3	50	33.33%

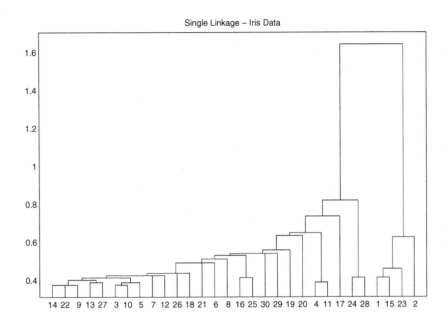

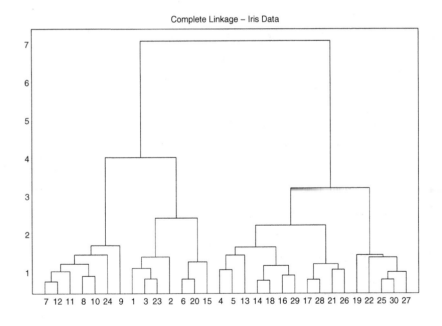

FIGURE 8.2
The top dendrogram is based on Euclidean distance and single linkage. There is some chaining in this case. The bottom dendrogram uses complete linkage and seems more reasonable.

```
tabulate(cidCL)
% This is shown in the window.
```

Value	Count	Percent
1	28	18.67%
2	50	33.33%
3	72	48.00%

We see in both sets of clusters that there is one group of size 50. This is likely the species of iris that is well-separated from the other two species. Single linkage and complete linkage provide different groupings for the other two species. Note that single linkage separates these two species into two groups, one with only two observations. Again, this is because of how single linkage measures closeness between clusters and can produce chaining. Complete linkage separates these two species differently.

We show the clusters from complete linkage in Figure 8.3. This is the code to create the plot.

```
% We can get a grouped scatter plot matrix.
% Use the complete linkage clusters.
varNames={'S-Length','S-Width','P-Length','P-Width'}
gplotmatrix(X,[],cidCL,[],...
      '.ox',3.5,'off','variable',varNames);
title('Clusters in Iris Data - Complete Linkage')
```

It is always a good idea to assess the clustering results using visualization, such as grouped scatter plots.

A brief discussion of the cluster labels is worthwhile. In the output tables above, we see that the cluster of size 50 has different labels. It is called cluster three in single linkage and cluster two in complete linkage. The cluster labels are arbitrary and do not directly connect with true class labels or ones from other clustering methods.

This example also illustrates our earlier point—that the groups we get from agglomerative clustering depend on the type of linkage used. We did not give an example of using other distance metrics, but the idea is the same. In summary, the clusters depend on the methods used to find them. We will see this also in our next clustering approach.

8.2.2 K-Means Clustering

K-means clustering uses an optimization criterion or objective function to partition the data into K groups. The value of K is specified in advance. This is in contrast to agglomerative clustering that finds a multilevel hierarchy of partitions, and the user can ask for any number of clusters (up to n) once the process is done. If we want a different number of groups in K-means, then we have to run the method again.

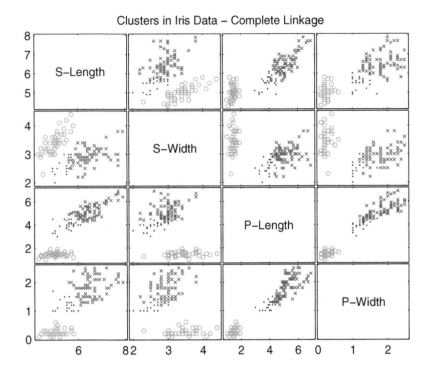

FIGURE 8.3

The group IDs in this scatter plot matrix are determined by agglomerative clustering using Euclidean distance and complete linkage. We see that one of the species of iris is well-separated from the other cloud of points. This is the one group of size 50 that we got from both single and complete linkage. The rest of the points are separated into two groups, as indicated by the different plotting symbols.

One of the advantages of an optimization based method like K-means is that it only needs the data as input. It does not require the $n \times (n - 1)/2$ unique interpoint distances that we needed with hierarchical approaches. This can be important when we are dealing with very large data sets.

The goal of K-means clustering is to partition the data into K disjoint groups, such that the within-group sum-of-squares is minimized. Methods that use this criterion tend to produce clusters with a hyperellipsoidal shape. The basic K-means procedure is presented here.

1. Find K initial cluster centroids. The starting points can be chosen randomly from the data or specified by the user.

2. Calculate the distance between each observation and each cluster centroid.

3. Place each data point into the group with the closest centroid.

4. Calculate or update the centroid (say the p-dimensional mean) of every cluster using the observations grouped there. These are the new cluster centers that are used in step 2.

5. Repeat from step 2 until no changes in cluster memberships are made.

One could get empty clusters with this method, but usually the procedure yields K groups, as requested.

This is an iterative method that depends on the starting point and distance metric used. Therefore, it is a good idea to explore the clusters one gets from K-means using different distances and starting points, as well as for other values of K.

MATLAB provides a function in the Statistics Toolbox called **kmeans**. We will explore its use in the next example. The default function call is

$$\texttt{cids = kmeans(X,K)}$$

where **X** is the data matrix, **K** is the number of clusters to return, and **cids** is the n-dimensional vector of cluster labels. As usual, the default is to use the squared Euclidean distance.

It is important to note that the type of distance used will determine how the centroid is calculated. We list the options for **kmeans** here.

- **Squared Euclidean distance** (**'sqEuclidean'**): This is the default. The centroid is the p-dimensional mean.

- **City block** (**'cityblock'**): This is the sum of the absolute differences. The centroid is the p-dimensional median of the points in that cluster.

- **Cosine** (**'cosine'**): This is calculated as one minus the cosine of the included angle between the data points. The centroid is found by first normalizing the observations to unit Euclidean length and then finding the p-dimensional mean.

- **Correlation** (**'correlation'**): This measure is found as one minus the correlation between observations. The points in a cluster are first transformed to have zero mean and unit standard deviation. The centroid is then calculated as the component-wise mean.

- **Hamming** (**'Hamming'**): Suitable only for binary data, this is found as the percentage of components that differ, and the centroid is given by the component-wise median.

To specify the distance, use the input name **'distance'** with one of the values listed above.

There are other name-value pairs that one can use to control the actions of kmeans. One useful one is **'emptyaction'**, which can be used to tell MATLAB what to do when a cluster becomes empty. The default is to return an error, but you could specify a value of **'singleton'** that will replace the empty cluster with a new starting point.

As we said previously, *K*-means uses an iterative algorithm. In these cases, it is good to have different random starting points with the goal of getting a more optimal solution. You can ask MATLAB to perform *K*-means for some given number of replicates or starting points. The initial centroids can be found using a random subsample from the data set (default), a selection of *K* points uniformly chosen from the range given by the data, or a preliminary clustering phase. The starting point for these replicates is specified using the **'start'** name argument, as we show in the next example.

Example of K-Means Clustering

We use the **iris** data to illustrate **kmeans** and to show how we can assess the cluster output. First, we make sure that the data are in the workspace, and we construct the data matrix by stacking all species of iris in a matrix **X**.

```
% Load the data
load iris

% Create a data matrix.
X = [setosa; versicolor; virginica];
```

Next, we ask for three clusters based on the cosine distance. We also specify a starting point of *K* centers that are randomly chosen from the range of **X**.

```
% Apply K-means clustering using the
% 'cosine' distance and a random start
% selected uniformly from the range of X.
cidK3 = kmeans(X,3,...
        'distance','cosine',...
        'start','uniform');
```

We can get some idea of the clusters by looking at the distribution of cluster IDs using the **tabulate** function.

```
tabulate(cidK3)

% This is shown in the window.
```

Value	Count	Percent
1	55	36.67%
2	45	30.00%
3	50	33.33%

We cannot directly map the cluster IDs to the true class labels, but this seems reasonable, just from the number of observations in each cluster. It looks better than the results we had from agglomerative clustering.

Example of a Silhouette Plot

There is a special type of plot that one can use with the output from cluster methods. This is called the *silhouette plot*, and it displays a measure of how close each data point is to observations in its own cluster as compared to observations in other clusters.

Kaufman and Rousseeuw [1990] developed a measure called the *silhouette width*. This is given for the *i*-th observation as

$$sw_i = \frac{b_i - a_i}{\max(a_i, b_i)},$$

where a_i is the average distance to all other points in its cluster. The b_i is found as follows. First, we find the distance between the *i*-th point and all points in another cluster c and take the average. This provides a measure of distance between the *i*-th point and cluster c. The minimum of these taken across clusters is b_i.

The silhouette width ranges from -1 to 1. If an observation has a silhouette width close to 1, then it is closer to observations in its own group than others. This is what we are seeking with clustering. If an observation has a negative silhouette width, then it is not well clustered.

A silhouette plot is essentially a type of bar chart that shows the silhouette width for the *n* observations grouped by cluster. We show how to get this plot for the clusters we obtained from *K*-means. Note that one can specify a different distance for the silhouette width, as we do here.

```
% Get a silhouette plot for K-means output.
silhouette(X,cidK3,'cosine');
title('Silhouette Plot - K-means')
```

The silhouette plot is shown in the top of Figure 8.4, and we see that the third cluster has values that are all close to one, indicating that the observations are nicely grouped. Groups one and two have some points with small values, and the first one has a few negative ones. This is an indication that these two groupings might be suspect.

The silhouette widths and silhouette plot can be used with cluster IDs from any clustering approach, and we now construct this plot for the groups we obtained from agglomerative clustering with complete linkage. The plot is displayed in the bottom of Figure 8.4. Cluster two seems well clustered, as we expected, but there are several negative silhouette values in the third group, which means that those points are not too similar to other observations in their cluster.

```
% Get a silhouette plot for the agglomerative
% clusters. That clustering used the default
% squared Euclidean distance.
```

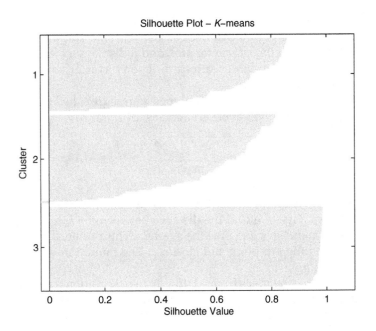

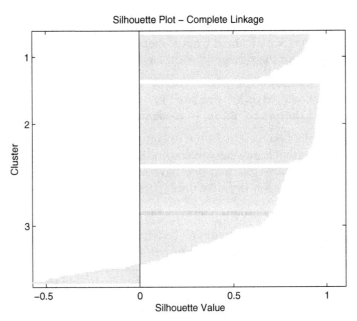

FIGURE 8.4
The display at the top is a silhouette plot for K-means clustering applied to the iris data. Silhouette values close to 1 indicate observations are grouped well. The bottom plot uses the cluster IDs from agglomerative clustering, and we see that this grouping is worse than K-means because there are many smaller and some negative silhouette widths.

```
silhouette(X,cidCL);
title('Silhouette Plot - Complete Linkage')
```

Comparing the two silhouette plots, it appears that the results from *K*-means are better than the agglomerative clusters.

We can also ask for the *n* silhouette values as output from the **silhouette** function. We could use these to get an overall measure of the groupings by taking the average.

```
% Get the average silhouette value of K-means clusters.
sK = silhouette(X,cidK3,'cosine');
msK = mean(sK)

msK =

    0.7491

% Get the average silhouette value of agglomerative
% clustering with complete linkage.
sC = silhouette(X,cidCL);
msC = mean(sC)

msC =

    0.6644
```

The average silhouette value for *K*-means is larger than the one we have for agglomerative clustering, which means it is a better cluster solution.

Example of K-Means Replicates

We conclude this section by showing how to request multiple runs of *K*-means. In this case, we are requesting four clusters based on the city block distance. We also specify six replicates or trials. The **'display'** option we chose will show the final results of every replicate.

```
% Get replicates to find the 'best' solution.
% Use the default starting point, 6 replicates,
% and display the results of each iteration.

cidK4 = kmeans(X,4,...
      'distance','cityblock',...
      'display','final',...
      'replicates',6);

Replicate 1, 7 iterations, total sum of distances = 137.8.
Replicate 2, 6 iterations, total sum of distances = 154.5.
Replicate 3, 3 iterations, total sum of distances = 136.5.
```

```
Replicate 4, 12 iterations, total sum of distances = 136.5.
Replicate 5, 6 iterations, total sum of distances = 150.4.
Replicate 6, 8 iterations, total sum of distances = 136.5.
Best total sum of distances = 136.5
```

This illustrates some of the points we made about *K*-means. In particular, it shows that the results depend on the starting point because we see different total sum of distances with the various iterations. The silhouette plot is given in Figure 8.5, and we see that it is a worse solution than the one we got from three clusters. This is also clear from the following average silhouette value, which is smaller that the value from our previous groupings.

```
% Get the average silhouette value.
sK4 = silhouette(X,cidK4,'cosine');
msK4 = mean(sK4)

msK4 =

    0.3913
```

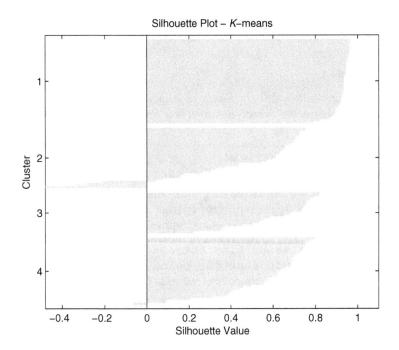

FIGURE 8.5
The silhouette plot for the 4-cluster solution from K-means is worse than the previous 3-cluster grouping using K-means (see Figure 8.4 top).

8.3 Summary and Further Reading

We presented two major approaches in machine learning in this chapter. The first had to do with ways we can perform supervised classification where we have class labels associated with each of our observations. The second area covered methods for unsupervised learning or clustering where we do not have group labels or we choose to explore the data without using them.

We covered the main techniques of classification, including linear and quadratic classifiers, naive Bayes classification, and the K-nearest neighbor classifier. There are other options in MATLAB that we briefly describe next, and we also list them in Table 8.1. One is a parametric approach that uses a generalized linear model to estimate the relationship between the features and the class label. The features act as predictors, and the class label is the response. For example, the two-class case corresponds to logistic regression, where we have two binary outcomes—0 and 1. We did not cover generalized linear models in this primer, but we provided some functions for building them in Chapter 6.

TABLE 8.1

Supervised Learning Functions in the Statistics Toolbox

Function Name	Description
`classify`	Discriminant analysis, not object-oriented
`crossval`	Cross-validation to estimate error
`fitcdiscr`	Create linear and quadratic objects
`fitcknn`	Build a K-nearest neighbor classifier
`fitcsvm`	Estimate a support vector machine classifier
`fitctree`	Build classification tree object
`fitensemble`	Build an ensemble classifier (bagging or boosting)
`fitNaiveBayes`	Construct a Naive Bayes classifier
`perfcurve`	Construct an ROC (Receiver Operating Characteristic) performance curve

We can think of a K-nearest neighbor classifier as nonparametric because we are not explicitly modeling the class-conditional probabilities with a parametric distribution, as we do with linear or quadratic classifiers. The Statistics Toolbox has other nonparametric methods for classification. These are classification trees, support vector machines, and ensemble methods. We briefly describe them here.

Classification and regression trees are a nonparametric approach to modeling the relationship between a response variable and predictors. The methods produce a binary decision tree, which is estimated from the data. To obtain a predicted response for a given set of predictors, one would start at the root node, making decisions at each node based on the predictors, until a leaf node is reached. The value in the leaf node is the estimated response. If the responses are nominal, such as group labels, then the tree is a *classification tree*. If the responses are numeric (non-categorical), then it is a *regression tree*.

The function used for fitting classification trees is called **fitctree**, and it produces a classification tree data object. We can apply many of the same methods from other classification objects, but it also has some additional ones that are unique to trees, such as **prune**, **compact**, and **edge**. You can view a list of methods for the classification tree object by displaying the tree at the command line and clicking on the *methods* hyperlink.

A *weak learner* is a classifier that has a prediction accuracy that is just slightly better than random guessing. The idea behind *ensemble methods* is to combine multiple weak learners to produce a better classifier. The single weak learners are typically developed by introducing some randomness into the process, perhaps through subsampling. Two commonly used approaches are *bagging* and *boosting*. The MATLAB function **fitensemble** provides several options for fitting ensemble classification and regression models.

The last type of classifier we mention can be used when we have just two classes. These are called *support vector machines*. They use an optimization approach to build a classifier, which is given by finding the best hypersurface that separates the observations of each class. The function **fitsvm** will train a support vector machine classifier and produce an **SVMModel** data object.

The next topic we covered in this chapter was clustering or unsupervised learning. We discussed two main approaches—agglomerative hierarchical clustering and *K*-means clustering. The MATLAB functions appropriate for clustering are summarized in Table 8.2.

TABLE 8.2

Unsupervised Learning Functions in MATLAB®

cophenet	Cophenetic correlation coefficient
cluster	Returns *K* clusters from agglomerative clustering
clusterdata	Complete agglomerative cluster solution
dendrogram	Dendrogram tree
evalclusters	Evaluate cluster solution
fitgmdist	Gaussian mixture approach
kmeans	*K*-means cluster solution
linkage	Hierarchical tree from interpoint distances
gmdistribution.fit	Gaussian mixture approach
pdist	Get interpoint distances
silhouette	Silhouette values and plot

Agglomerative clustering constructs a set of nested partitions or groups of the data based on the interpoint distance metric and the type of linkage method used to measure how close two groups are to one another. Changing these can sometimes yield different clusters. The number of clusters is specified after the method finishes, and the choice for the number of groups can be determined by looking at a dendrogram plot.

K-means clustering is a method for clustering that optimizes an objective function. We presented two ways we can assess the result of K-means and other clusterings. These were the silhouette plot and the average silhouette width. There is another statistic we can use with hierarchical clustering. This is called the ***cophenetic correlation coefficient***, and it measures how well the tree (or dendrogram) represents the interpoint distances. A value close to one indicates a good clustering solution. The MATLAB function **cophenet** will calculate the cophenetic correlation coefficient between the linkage given in **Z** and the interpoint distances in **Y**.

There is a third method for clustering that is implemented in MATLAB. This one uses a probability density estimation approach to clustering. Here, we assume that the data can be modeled as a finite mixture or weighted sum of K multivariate normals. Each of the components corresponds to a group or cluster. Observations are assigned to the component in the mixture with the highest posterior probability, similar to discriminant analysis. The function **gmdistribution.fit** returns a cluster solution given the data and a value for K.

There are many good references to help you learn more about classification and clustering, and we list a few of them here. A text that we recommend highly is *Pattern Classification* by Duda, Hart, and Stork [2001]. This book is an update of the 1973 text [Duda and Hart, 1977], and it includes references and examples using MATLAB. Another good book is *Pattern Recognition and Machine Learning* by Bishop [2006]. Hastie, Tibshirani, and Friedman [2009] wrote an excellent text on machine learning that provides more information about classification and clustering. For a book that focuses only on clustering, we recommend *Cluster Analysis* by Everitt, Landau, and Leese [2001].

The *Computational Handbook with MATLAB* [Martinez and Martinez, 2007] has a chapter on statistical pattern recognition that includes sections on classification and clustering. The *Exploratory Data Analysis with MATLAB* text [Martinez et al., 2010] has two chapters on clustering, one of which is on model-based clustering. Model-based clustering uses density estimation to form groups [Fraley and Raftery, 1998].

References

Anderson, E. 1935. "The irises of the Gaspe Peninsula," *Bulletin of the American Iris Society*, **59**:2–5.

Andrews, D. F. 1972. "Plots of high-dimensional data," *Biometrics*, **28**:125–136.

Bertin, J. 1983. *Semiology of Graphics: Diagrams, Networks, Maps*, Madison, WI: The University of Wisconsin Press.

Bhattacharyya, G. K. and R. A. Johnson. 1977. *Statistical Concepts and Methods*, New York: John Wiley and Sons.

Bishop, C. M. 2006. *Pattern Recognition and Machine Learning*, New York: Springer.

Borg, I. and P. J. F. Groenen. 2005. *Modern Multidimensional Scaling: Theory and Applications, 2nd Edition*, New York: Springer.

Box, G. E. P. and N. R. Draper. 1987. *Empirical Model-Building and Response Surfaces*, New York: John Wiley and Sons.

Cattell, R. B. 1966. "The scree test for the number of factors," *Journal of Multivariate Behavioral Research*, **1**:245–276.

Chambers, J. M., W. S. Cleveland, B. Kleiner, and P. A. Tukey. 1983. *Graphical Methods for Data Analysis*, Boston, MA: Duxbury Press.

Cleveland, W. S. 1993. *Visualizing Data*, New York: Hobart Press.

Cook, D. and D. F. Swayne. 2007. *Interactive and Dynamic Graphics for Data Analysis: With R and GGobi (User R)*, New York: Springer–Verlag.

Cook, R. D. 1998. *Regression Graphics: Ideas for Studying Regressions through Graphics*, New York: John Wiley & Sons.

Cormack, R. M. 1971. "A review of classification," *Journal of the Royal Statistical Society, Series A*, **134**:321–367.

Cox, T. F. and M. A. A. Cox. 2001. *Multidimensional Scaling, 2nd Edition*, London: Chapman and Hall/CRC.

Davis, T. A. 2010. *MATLAB Primer*, Boca Raton: CRC Press.

Devore, J. L. and K. N. Berk. 2012. *Modern Mathematical Statistics with Applications*, New York: Springer–Verlag.

Draper, N. R. and H. Smith. 1981. *Applied Regression Analysis, 2nd Edition*, New York: John Wiley & Sons.

Duda, R. O. and P. E. Hart. 1977. *Pattern Classification and Scene Analysis*, New York: John Wiley & Sons.

Duda, R. O., P. E. Hart, and D. G. Stork. 2001. *Pattern Classification, Second Edition*, New York: John Wiley & Sons.

Efron, B. and R. J. Tibshirani. 1986. "Bootstrap methods for standard errors, confidence intervals, and other measures of statistical accuracy," *Statistical Science*, 1986, 1:54–77, http://projecteuclid.org/euclid.ss/1177013815, last accessed September 2014.

Efron, B. and R. J. Tibshirani. 1993. *An Introduction to the Bootstrap*, London: Chapman and Hall.

Everitt, B. S., S. Landau, and M. Leese. 2001. *Cluster Analysis, Fourth Edition*, New York: Edward Arnold Publishing.

Fisher, R.A. 1936. "The use of multiple measurements in taxonomic problems," *Annals of Eugenics*, **7**:179–184.

Fraley, C. and A. E. Raftery. 1998. "How many clusters? Which clustering method? Answers via model–based cluster analysis," *The Computer Journal*, **41**:578–588.

Gabriel, K. R. 1971. "The biplot graphic display of matrices with application to principal component analysis," *Biometrika*, **58**:453–467.

Glantz, S. and B. Slinker. 2001. *Primer of Applied Regression & Analysis of Variance*, New York: McGraw Hill.

Gower, J. C. 1966. "Some distance properties of latent root and vector methods in multivariate analysis," *Biometrika*, **53**:325–338.

Hanselman, D. C. and B. L. Littlefield. 2011. *Mastering MATLAB*, New Jersey: Prentice Hall.

Hartigan, J. A. 1967. "Representation of similarity measures by trees," *Journal of the American Statistical Association*, **62**:1140–1158.

Hastie, T., R. Tibshirani, and J. Friedman. 2009. *The Elements of Statistical Learning: Data Mining, Inference, and Prediction*, New York: Springer-Verlag.

Hyndman, Rob J. and Y. Fan. 1996. "Sample quantiles in statistical packages," *The American Statistician*, **50**:361-265.

Inselberg, A. 1985. "The plane with parallel coordinates," *The Visual Computer*, **1**:69–91.

Jackson, J. E. 1991. *A User's Guide to Principal Components*, New York: John Wiley & Sons.

Joliffe, I. T. 1986. *Principal Component Analysis*, New York: Springer–Verlag.

Kaufman, L. and P. J. Rousseeuw. 1990. *Finding Groups in Data: An Introduction to Cluster Analysis*, New York: John Wiley & Sons.

Kay, Stephen. 2005. *Intuitive Probability and Random Processes Using MATLAB*, New York: Springer–Verlag.

Kotz, S. and N. L. Johnson (Editors-in-Chief). 1986. *Encyclopedia of Statistical Sciences, volume 7*, New York: Wiley–Interscience.

Kruskal, J. B. and M. Wish. 1978. *Multidimensional Scaling*, Sage University Paper Series on Quantitative Applications in the Social Sciences, Thousand Oaks, CA: Sage.

Lilliefors, H. W. 1967. "On the Kolmogorov-Smirnov test for normality with mean and variance unknown," *Journal of the American Statistical Association*, **62**:399–402.

Marchand, P. and O.T. Holland. 2002. Graphics and GUIs with MATLAB, Third Edition, Boca Raton: CRC Press.

Martinez, W. and A. R. Martinez, 2007, *Computational Statistics Handbook with MAT-LAB, 2nd Edition*, Boca Raton: CRC Press.

Martinez, W., A. R. Martinez, and J. L. Solka. 2010. *Exploratory Data Analysis with MATLAB, 2nd Edition*, Boca Raton: CRC Press.

Robbins, N. B. 2005. *Creating More Effective Graphs*, New York: John Wiley & Sons.

Ross, S. 2012. *A First Course in Probability, 9th Edition*, New York: Pearson.

Sahai, H. and M. I. Ageel. 2000. *The Analysis of Variance: Fixed, Random and Mixed Models*, Boston, MA: Birkhauser.

Scott, D. W. 1992. *Multivariate Density Estimation: Theory, Practice, and Visualization*, New York: John Wiley & Sons.

Silverman, B. W. 1986. *Density Estimation for Statistics and Data Analysis*, London: Chapman and Hall/CRC.

Stigler, S. M. 1977. "Do robust estimators work with real data?" *Annals of Statistics*, **5**:1055–1078.

Strang, G. 1993. *Introduction to Linear Algebra*, Wellesley, MA: Wellesley-Cambridge Press.

Theus, M. and S. Urbanek. 2009. *Interactive Graphics for Data Analysis: Principles and Examples*, Boca Raton: CRC Press.

Tipping, M. E. and C. M. Bishop. 1999. "Probabilistic principal component analysis," *Journal of the Royal Statistical Society, Series B (Statistical Methodology)*, **61**:611–622.

Torgerson, W. S. 1952. "Multidimensional scaling: 1. Theory and method," *Psychometrika*, **17**:401–419.

Tukey, John W. 1977. *Exploratory Data Analysis*, New York: Addison-Wesley.

Turner, J. R. and J. F. Thayer. 2001. *Introduction to Analysis of Variance: Design, Analysis & Interpretation*, Thousand Oaks, CA: Sage Publications.

Unwin, A., M. Theus, H. Hofmann. 2006. *Graphics of Large Datasets: Visualizing a Million*, New York: Springer-Verlag.

Wainer, H. 1997. *Visual Revelations: Graphical Tales of Fate and Deception from Napoleon Bonaparte to Ross Perot*, New York: Copernicus/Springer–Verlag.

Wainer, H. 2004. *Graphic Discovery: A Trout in the Milk and Other Visual Adventures*, Princeton, NJ: Princeton University Press.

Wegman, E. J. 1990. "Hyperdimensional data analysis using parallel coordinates," *Journal of the American Statistical Association*, **85**:664–675.

Wilk, M. B. and R. Gnanadesikan. 1968. "Probability plotting methods for the analysis of data," *Biometrika*, **55**:1–17.

Wilkinson, L. 1990. *SYSTAT: The System for Statistics*, Evanston, IL: SYSTAT, Inc.

Wish, M. 1971. "Individual differences in perceptions and preferences among nations," in *Attitude research reaches new heights*, C. W. King and D. Tigert, eds., Chicago: American Marketing Association, 312–328.

Index of MATLAB® Functions

Subject Index

Printed in the United States
by Baker & Taylor Publisher Services